AF789716

STEPHEN MOYLAN

Muster-Master General
Secretary and Aide-de-Camp to Washington

BY

MARTIN I. J. GRIFFIN

NEW YORK:
GARRETT PRESS, INC.
1969

SBN 512-00280-0
LIBRARY OF CONGRESS CATALOG CARD
NUMBER 9-8923

This is a photographic reprint of the 1909 edition,
published in Philadelphia by Martin I.J. Griffin.
Reproduced from a copy in the Garrett Press Collection.

First Garrett Press Edition Published 1969

Manufactured in the United States of America

E
207
.B9 G8
1969

218978

GARRETT PRESS, INC.
Publishers
250 West 54th Street, New York, N.Y. 10019

A MOYLAN DRAGOON

STEPHEN MOYLAN

Muster-Master General
Secretary and Aide-de-Camp to Washington
Quartermaster-General
Colonel of Fourth Pennsylvania Light Dragoons and
Brigadier-General of the War for American Independence

The First and the Last President of the
Friendly Sons of St. Patrick of Philadelphia

BY

MARTIN I. J. GRIFFIN

PHILADELPHIA
1909

COPYRIGHT, JANUARY 16, 1909
BY
MARTIN I. J. GRIFFIN

STEPHEN MOYLAN, MUSTER-MASTER GENERAL, SECRETARY AND AIDE-DE-CAMP TO WASHINGTON, QUARTERMASTER-GENERAL, COLONEL OF 4TH PENNSYLVANIA LIGHT DRAGOONS AND BRIGADIER-GENERAL OF THE WAR FOR AMERICAN INDEPENDENCE—THE FIRST AND THE LAST PRESIDENT OF THE FRIENDLY SONS OF ST. PATRICK OF PHILADELPHIA.

"Shall we never leave off debating and boldly declare Independence?" [Moylan to Reed, Jan. 30, 1776.]

"America will—it must—be free." [Moylan to Robert Morris after Battle of Princeton.]

"I entered the service in the first year of the war, with a firm determination of prosecuting it to the end. I made up my mind, and my affairs for that purpose. I have shared its fatigues, its dangers and its pleasures with Your Excellency ever since—a man who has sacrificed everything for the service of his country." [Moylan to Washington, December 15, 1782.]

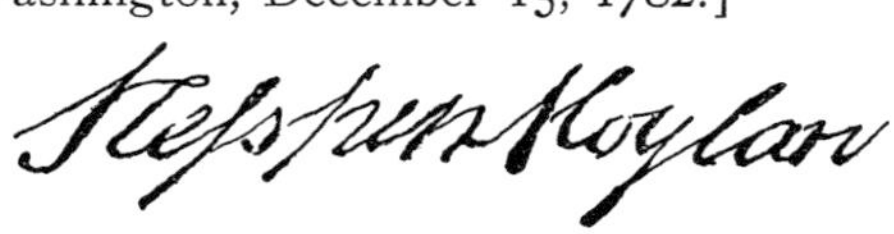

THE MOYLAN FAMILY.

General Stephen Moylan, of the American Revolutionary Army, was born in 1737 at Cork, Ireland.

According to Campbell's *History of the Hibernian Society and the Friendly Sons of St. Patrick,* Stephen Moylan was the son of John Moylan and the Countess of Limerick.

The Letters of Abbé Edgeworth and Memoir of his Life, by Father England, pastor of Passage near Cork, brother of Bishop England of Charleston, South Carolina, says John Moylan was "extensively engaged in mercantile pursuits."

He died in 1799, probably in Dublin. An abstract of his will, on record at the Four Courts, Dublin, relates:

John Moylan, of the city of Cork, merchant, being now on the eve of my departure for England, do make this my last will and testament.

To wife, Mary Moylan, the entire of my property in money, goods, debts or lands, in confidence that she will dispose of it amongst my children.

To David Andre, Esq., of London (as a small compensation for the heavy loss he formerly sustained by me and my brother), the sum of £500.

" I lament that I have it not in my power to leave my children in better circumstances, but my own misfortunes were great and heavy for a number of years, and it was only within these last three years that it pleased God to give me some little success."

Sole executrix, Said Wife.

Dated 28 June, 1797.

JOHN MOYLAN.

Declaration of Stephen Roche, Jun., of North Great George's Street in the city of Dublin, Esq. Sworn 12 Nov., 1799.

Proved 13 Nov., 1799, by Mary Moylan, widow of testator and sole executrix named in will.

Campbell states that the children of John Moylan and the Countess were Stephen, Francis and two daughters who became Ursuline Nuns. By his second wife, Mary, he had issue Jasper and John.

On a recent visit to Cork the compiler discovered the granddaughter of another son, Richard. There was also a son named James, as General Moylan in writing to General Washington spoke of a James Moylan as " a brother." He was a resident of Philadelphia in 1771-2 and during the Revolution was Agent of the United States in France in partnership with Gourlade in furnishing supplies to American cruisers. This is sustained by the Letters of Abbé Edgeworth saying that two sons of John Moylan " at an early age emigrated to America and served with success the American colonies as general officers in the eventful contest which terminated in the freedom of the United States."

At the time of the birth of Stephen, as well as that of his brothers and sisters, the Catholic religion in Ireland was under penal-law restrictions and penalties. Yet the Moylan family were so strong-hearted in the principles of faith and piety that Francis became a priest and subsequently Bishop of Cork, and the two daughters became Ursuline Nuns.

Mary Moylan—Sister Mary Aloysius—was born July 29, 1753, and entered the novitiate of the Ursulines of Cork on December 25, 1771; was professed April 26, 1774. During her life, a long one, she filled with great credit the principal posts in the Convent and was much loved and respected by the community and pupils. She lived to the advanced age of ninety, dying April 26, 1842.

Her sister, Miss B. Moylan, joined the Ursulines July 2, 1780; was professed January 13, 1783, and after several years of ill health died in October, 1842. Her name in religion was Sister Mary John Evangelist. [Records of the Convent at Blackrock, Ireland.]

There was also a daughter Anne, the residuary legatee of her brother the Bishop. This would make eight children of John Moylan.

"The Moylans were merchants, established in business probably as early as 1720. Dennis Moylan, the uncle of Stephen, who died in 1772, held the government contract for the commissariat of the Isle of Bourbon." [Campbell.]

His name was probably David.

Another uncle, Rev. Patrick Doran, a Jesuit, died in Cork in 1771-2 and was buried in the family burial lot in Upper Shandon. [Foley's *Records,* S. J., VII, p. 81.]

The penal laws against Catholics debarred their education in Ireland:

"If any Papist shall publicly teach school or instruct youth in learning in any private house, or shall be entertained to instruct youth, as usher or assistant to any Protestant schoolmaster, he shall be esteemed a Popish regular clergyman and prosecuted as such, and shall incur such penalties and forfeitures as any Popish regular convict is liable unto."

Even the sending of youth abroad to be educated was a penal offense. Yet Stephen and Francis "were smuggled out of the country to France for their education," says Campbell. It is probable that, as the Moylan brothers were engaged in commerce at Lisbon, that it was in that city of Portugal they were educated, and that there Stephen entered into business before coming to America. But as Stephen at one time desired to represent the United States in Spain, it is possible, however, that he may have been educated in that country.

At Lisbon, Portugal, in 1765 he was associated in business with David Moylan and Edward Forrest. The Registry of Vessels at the Port of Philadelphia for that year records the brigantine "John and David," of one hundred tons, built at Philadelphia, as being owned by John and David Moylan of Cork and Messrs. David and Stephen Moylan and Edward Forrest, British subjects residing at Lisbon. George Gould was Master.

John, of Cork, was the father of Stephen, of Lisbon. David, of Cork, was the brother of John and uncle of Stephen. David, of Lisbon, was, doubtless, the son of David, of Cork. So that the fathers in Cork were in business with their sons in Lisbon.

General Moylan's brother Francis, born September 17, 1735, was consecrated Bishop of Kerry in 1774, and died February 10, 1815. In 1786 he was translated to Cork to succeed Bishop John Butler who, at the age of seventy years, had apostatized and married, to inherit the estate of the Earl of Dunboyne. On August 22, 1787, in the Protestant Church of Clonmel he "read his recantation of the errors of the Church of Rome." He died May 8, 1800, repentant, bequeathing property to the Catholic College at Maynooth.

Bishop Moylan was educated at Paris and afterwards at the University of Toulouse, where he studied Theology. He was buried in a vault in his cathedral.

References—Short Life of Dr. Moylan in an appendix to Hutch's Life of Nano Nagle, Letters from the Abbé Edgeworth to his Friends with Memoirs of his Life, including some account of Dr. Moylan by the Rev. T. R. England, Fitzpatrick's Irish Wits and Worthies, Fitzpatrick's Secret Service under Pitt, Castlereagh Papers, Sarah Atkinson's Life of Mary Aikenhead, Husenbeth's Life of Dr. Milner, O'Renehan's Collections on Irish Church History, Caulfield's Council Book of the Corporation of the City of Cork, History of Nat. Biog.

His will sets forth:

Francis Moylan, Roman Catholic Bishop of Cork. To the Rev. John Murphy, Rev. Jeremiah Collins and Rev. John England of this city, priests, all the government debentures deposited in the hands of Messrs John Roche & Co., Dublin, by Messrs Stephen & James Roche of this city, Bankers, for my account. I also bequeath to them Pierse & Christina's bond of £400 with interest . . . in trust. . . . I appoint my sister Anne my residuary legatee. Executors, said trustees.

D. 13 April, 1814.

F. MOYLAN.

Description of Property.

To be interred in the vault of this chapel.

Codicil d. 13 April, 1814.

2d codicil. . . .

Will & two codicils proved 12 Sept., 1815, by Rev. John England, one of the executors.

Stephen Moylan " received a good education, resided in England and then came to America, where he traveled extensively and finally settled in Philadelphia," says Appleton's *Cyclopædia of American Biography.* [IV, 1888.]

STEPHEN MOYLAN IN AMERICA.

He came to Philadelphia in 1768. Though then but a young man of about thirty he must have been in possession of means to warrant his being received into the social element of the chief city of the Province and to have had the education and polite accomplishments justifying his entry and association with the leading men of the city and to have found his Faith no debarment of the recognition due a gentleman.

He engaged in commerce. On November 11, 1768, the brigantine " Richard Penn," of 30 tons, built at Philadelphia, James Galoway, Master, was registered as owned by Stephen Moylan of Philadelphia.

The following April 7, 1769, the brigantine " Minerva," George Barwick, Master, 120 tons, is also registered. Stephen Moylan, owner.

An advertisement dated March 23, 1769, by William Kelley, offers for sale or " lett," " two thousand acres in Morris County, New Jersey, as healthy a country as any in the world," fifteen miles from Newark, " a sea port town," and twenty-three miles from New York. Among those of whom inquiries might be made was Stephen Moylan, Esq., of Philadelphia. [*Pa. Gaz.,* May 18th: Supplement No. 2108.]

The affix "Esq." had, in those days, more significance than in our times, when indeed it may be said it has none, so commonly is it applied. But in Colonial days it was a title of import, indicating a social condition and position distinguishing the one to whom it was applied as one of character and standing in the community.

On October 27, 1769, was registered the snow "Ceres" of 100 tons, built at Philadelphia and owned by Edward Forrest, Andrew Morrogh, Patrick James Morrogh, Dennis Conwell, Andrew Morrony, British subjects residing at Lisbon, and Stephen Moylan and Nicholas Bodkin both of Philadelphia. Bodkin was the Master of the vessel.

On August 27, 1770, the sloop "Santa Maria," built at Kingston, Province of Massachusetts Bay, 70 tons, Thomas Bell, Master, is registered as being owned by John Kennedy, a British merchant residing in the Island of Porto Rico and by Messrs. Willing and Morris, Stephen Moylan and William Marshall, all of Philadelphia.

On October 4, 1770, the ship "Don Carlos," a British-built vessel rebuilt at Cork, of 100 tons, Terrence Connor, Master, is registered as owned by Edward Forrest, a British subject residing at Lisbon, John and David Moylan of Cork and Stephen Moylan of Philadelphia.

But Captain Terrence Connor must not long have commanded the ship, for the following April (1771) he is registered as owner and Master of the schooner "Don Carlos," of thirty-five tons, built in the Province of Massachusetts Bay.

In 1770 Mr. Moylan became a member of the Gloucester Fox Hunting Club. This is another indication of his social status among the well-born and well-to-do people of the Province. We find this much more clearly shown by an entry in the diary of John Adams while attending the first Continental Congress when, on September 24, 1774, he records having "dined with Richard Penn; a magnificent house and most splendid feast and a very large company; Mr. Dickinson and General Lee were there and Mr. Moylan, besides a great number of the Delegates."

The distinction of "Mr. Moylan" must have been high amid the "very large company" to have warranted such a noting, especially when associated in mind with two such celebrities of that day as John Dickinson, author of "The Farmer's Letters," and General Charles Lee who, next year, was appointed second to Washington in command of the American Army raised to resist Great Britain.

THE FRIENDLY SONS OF ST. PATRICK.

On St. Patrick's Day, 1771, was formed the Friendly Sons of St. Patrick. At its organization it had twenty-four regular members and six honorary members. It was composed mainly of "prosperous merchants," many of them engaged in the shipping and importing business and dealing in European and East India goods." [Campbell.]

Stephen Moylan was elected the first President of The Friendly Sons of St. Patrick, though but two of his fellow-members, Thomas FitzSimons and George Meade were Catholics. This shows how free of religious prejudice were the founders of this most patriotic and worthy society. That spirit of liberality and justice ever pervaded The Sons of St. Patrick during its existence and is perpetuated in its successor, if not heir, the present Friendly Sons of St. Patrick for the Relief of Immigrants from Ireland.

Stephen Moylan served as President until June 17, 1773, when John M. Nesbitt succeeded, as Mr. Moylan was reported as "beyond sea," as were also Thomas FitzSimons and Ulysses Lynch.

At the St. Patrick's Day dinner of 1775 FitzSimons was present but Moylan was fined *7s 6d* for absence. This indicates he was in the city. Moylan appears not to have been present afterwards until the December meeting in 1781, but to have been "at camp" from 1775. He was present on St. Patrick's Day, 1782, but at June, 1782, and March, 1783, was at "camp." At September, 1783, he was "beyond sea," but present on St. Patrick's Day, 1784, when also Commodore John Barry, so long absent, also attended. On St. Patrick's Day, 1786, he is recorded present as "General Moylan." Ten years later he was elected President and Thomas FitzSimons Vice-President. No further records have been discovered. Associating, as we have seen, with the Delegates to the First Continental Congress and moving in the select social circles of the principal city of the Province, the foremost characters of the day called him "Friend," as did the author of the celebrated "Farmer's Letters," John Dickinson.

The Second Congress of the Continent met at the State House, Philadelphia, on May 10, 1775. The following month, Delegate George Washington, of Virginia, was elected Commander-in-Chief of all the forces besieging the British at Boston.

THE REVOLUTION.

The Revolution was on. War existed. The Colonies had not only resisted and fought the armed forces of England but had concentrated their resistance and chosen a Chief to command the disjointed forces which had battled with the British army and were assembled around Boston. Moylan was aroused and would add his endeavors to those battling for Liberty. "He desired to place himself in the line of usefulness for his adopted country." So from his friend John Dickinson he obtained this letter of introduction to Washington, then at camp at Cambridge, Massachusetts:

JOHN DICKINSON INTRODUCES MOYLAN TO WASHINGTON.

Dear Sir:—Mr. Moylan, a friend of mine, informs me that he intends to enter into the American Army. As he resided some years in this City and was much esteemed here, I sincerely hope he will be so happy as to recommend himself to your favour, which I am convinc'd he will endeavour to deserve.

I heartily wish you every kind of Happiness and am, Sir, Your Most Obedient Servant, JOHN DICKINSON.

Philadelphia,
July 25, 1775.
General Washington.

Endorsed by Washington: "From Jno Dickinson, Esq., 25th July, 1775."

[Potter's *Monthly,* Vol. VI, p. 14, 1876.]

To the letter Washington replied:

CAMP AT CAMBRIDGE, Aug. 30, '75.

Dear Sir:—Your favour of 25 ult. recommendatory of Mr. Moylan came duly to hand and I have the pleasure to inform you that he is now appointed Commissary-General of Musters—one of the offices which the Congress was pleased to leave at my disposal. I have no doubt, from your account of this Gentleman, of his discharging the duty with honour and fidelity.

For the occurrences of the Camp, I refer to my publick letters, address'd to Mr. Hancock, and am, with sincere regard,

Dr. Sir, Yr. Most Obedt Hble Servt,

To John Dickinson, Esq., Go. WASHINGTON.
Philadelphia.

[Dawson's *His. Mag.,* Aug., 1859, p. 243.]

APPOINTED MUSTER-MASTER.

On August 11, 1775, Washington had issued this order:

"The Commander-in-Chief has been pleased to appoint Stephen Moylan, Esq., to be Muster-Master General to the Army of the United Colonies. He is in all things touching his duty as Muster-Master General, to be considered and obeyed as such."

On the 14th Washington appointed Major Thomas Mifflin Quartermaster-General, and ordered:

"As the troops are all to be mustered as soon as possible, the Muster-Master General, Stephen Moylan, Esq., will deliver the commanding officer of each regiment thirty blank muster rolls, upon Friday next, and directions for each Captain how he is to fill up the blanks." [*Am. Ar.,* 4th Series, Vol. III, p. 250.]

CONDITIONS OF THE ARMY.

The condition of the army at the time Moylan was appointed Muster-Master General is set forth by Moylan's associate, Colonel Joseph Reed, in a letter to Elias Boudinot dated "Camp at Cambridge, August 13, 1775":

"We heard, and we find it true, that the Army was a scene of disorder and confusion, that the Officers were not only ignorant and litigious but scandalously disobedient, and in the last action many of them proved such notorious cowards that the very existence of the army, and consequently the salvation of America, depended upon an immediate reform. This could never have been made among themselves. It required all the weight and influence of General Washington, under a Continental commission and assisted by every one around him in whom he could confide, to execute this necessary work. . . . Such a scene opened to the General that I assure you there was reason to fear his supporting himself under it, if those of us who accompanied him had not pledged ourselves to give him every aid in our power. He expressed himself to me in such terms that I thought myself bound by every tie of duty and honour to comply with his request to help him through this sea of difficulties. The men who compose the army are tractable and generally well behaved, but by suffering them to choose their own officers it seems to me they have excluded every gentleman and have picked out such as would give them every indulgence or sought the service for the profit." [*MS.*]

Washington, writing, on August 29, 1775, to Richard Henry Lee, said:

"As we have now nearly completed our lines of defence, we have nothing more in my opinion to fear from the Enemy, providing we can keep our men to their duty and make them watchful and vigilant; but it is among the most difficult tasks I ever undertook in my life to induce the people to believe that there is or can be danger till the Bayonet is pushed at their breasts; not that it proceeds from any uncommon prowess, but rather from an unaccountable kind of stupidity in the lower class of these people, which believe me prevails but too generally among the officers of the Massachusetts part of the Army, who are nearly of the same kidney with privates. . . .

"There is no such thing as getting officers of this stamp to exert themselves in carrying orders into execution—to curry favour with men (by whom they were chosen, and on whose smiles possibly they may think they may again rely) seems to be one of the principal objects of their attention. . . .

"On Saturday night last we took possession of a spot within point blank shot of the enemy on Charlestown neck, worked incessantly the whole night with 1,200 men, and threw up an intrenchment such as to bid defiance to their cannon; about nine o'clock on Sunday they began a heavy cannonade which continued through the day without any injury to our work, and with the loss of four men only, two of which were killed. The cannonade however we were twice obliged to submit to with impunity not daring to make use of artillery on acct. of the consumption of powder, except with one nine pounder placed on a point, with which we silenced, & indeed sunk, one of their Floating Batteries.

"This move of ours, was made to prevent the Enemy from gaining this hill and we thought was giving them a fair challenge to dispute it for we had been told by various people who had just left Boston, that they were preparing to come out, but instead of accepting of it we learn that it has thrown them into great consternation. Yesterday afternoon they began a Bombardment without any effect, as yet." [*MS.*]

MOYLAN FITS OUT ARMED VESSELS.

In October, 1775. Congress learned that there had sailed from England without convoy, "two north country built brigs of no force loaded with arms, powder and other stores for Quebeck which it being of importance to intercept," General Washington, on October 4th, was directed to "apply to the Council of Massachusetts Bay for two armed vessels in their service and despatch the same with sufficient number of people, stores, &c., particularly a number of oars, in order, if possible, to intercept the said two brigs and their cargoes, and to secure the same for the use of the Continent; also any other transports laden with ammunition, clothing or other stores, for the use of the ministerial army or Navy in America, and to secure them in the most convenient places for the purpose above mentioned; . . . that the General be directed to employ the said vessels and others, if he judge necessary, to effect the purpose and that he be informed that Rhode Island and Connecticut vessels of force will be sent directly after them, to their assistance."

Though these were "privateers," as Washington spoke of them, the fitting out of these armed vessels may be said to be the beginning of the Navy of the United Colonies. It was designed not to attack British armed vessels but to intercept unarmed supply vessels without force so as to capture the supplies going to Quebeck to the British forces there, and later those going to Boston where the British were besieged.

It was with the fitting out of the two and, later, seven armed vessels that the Muster-Master General coöperated with Colonel John Glover of the Marblehead "Marine" Regiment composed principally of seafaring men.

On October 4, 1775, Washington appointed Colonel John Glover and Muster-Master General Moylan to fit out two "prime sailers, to put them in the best order and lose no time." So Colonel Joseph Reed wrote, by direction of General Washington, to Colonel Glover, at Marblehead on October 4, 1775, adding:

"Mr. Moylan, the Muster-Master General is associated with you in this business; and whatever engagements are entered into by you and Mr. Moylan, when you may happen to be together, or by either in case one goes to Newbury, the General will fully ratify and confirm."

"Instructions" were also sent as to the kind of vessels, their appraisements, armament, etc.

On the 7th of October Washington wrote Moylan at Salem or Marblehead that if he thought he could, without inconvenience, go to Portsmouth to conduct matters relative to a supply of flour to do so and settle the matter on the best terms.

Two days later Moylan and Glover wrote Washington from Salem that owing to difficulty of procuring carpenters it would be Saturday before the first vessel would be ready, and on that day the Captain and Company might be sent. On the following Wednesday the other vessel would be ready.

They related the difficulties which occurred in the hiring of the vessels; that a schooner had been hired from Mr. Stevens of Marblehead noted for her good qualities and would be ready in twelve or fourteen days; that yesterday two large ships of war were seen coming out of Boston harbor, which it was believed were going to Portsmouth. "Mr. Moylan will set out for that place to-morrow."

"Your Excellency may be assured we have used our best intelligence in transacting this business and will continue to do so in fitting them out, appointing agents and in every command you may please to honour us with."

On October 11, 1775, Colonel Reed, for Washington, replied that it was "a disappointment that the vessels cannot be got ready sooner," as a "number of transports from England are hourly expected on the coast." So Washington directed: "Not a moment of time be lost in getting them ready and proceed to Newbury and take up a fourth vessel on the same service. Should Mr. Moylan be gone to Portsmouth, Colonel Glover was to forward this letter by express that he may return to Newburyport to take up the fourth vessel and let us know what he will want to equip the vessel for sea."

The vessels of Captain Nicholas Broughton and Captain John Selman were engaged and fitted out. On October 12th Colonel Reed notified both: "Lose no time. Everything depends upon expedition."

Broughton was appointed Captain and directed to "proceed on board the schooner 'Hannah' at Beverly" and to "cruise against such vessels as may be found on the high seas in the service of the Ministerial Army and to seize all such vessels laden with soldiers, arms, ammunition or provisions."

On October 10, 1775, Moylan was at Newbury, Mass., and there appointed Tristam Dalton, Agent, to take care of any prizes that might be sent in by the three armed schooners fitting out to distress the enemy in Boston.

From Newburg Moylan went to Portsmouth where he arrived on 11th. He delivered to the Chairman of the Committee of Safety Colonel Reed's letter, with the result that the Committee agreed to deliver to Moylan 1200 barrels of flour and would let him know what would be done with the rest of the cargo when they heard from General Washington, to whom they had written, "but," reported Moylan to Washington, "as a half loaf is better than no bread, I told them I would take immediate charge thereof, and yesterday I engaged two sloops to begin the transportation."

Concerning the flour he wrote: "I find by having had a few barrels weighed they fall short 3, 4 & 5 pounds of the weight marked on them, and as it will probably become a Continental charge I have thought it best to have them all weighed, that the publick, or the commissary into whose hands they may fall, may not pay for more flour than they really have. As there are people in this town inimical to the Cause, I think it best (in part) to keep it secret where the flour is to be landed. The engagements I have made are that they must proceed to Newburg, Ipswich, Cape Ann, Salem or Marblehead according to the orders I shall give them when going to sea."

He reported that he had appointed Tristram Dalton Agent for the prizes and also Colonel Joshua Wentworth for the Province, who are "to obey all orders received from Headquarters."

He requested Washington to send him two hundred and forty dollars to pay freight and other charges.

To this report came the reply: "The General is pleased with your proceedings." Colonel Reed added: "We are very anxious to hear of the armed vessels being ready for sea. Every day, every hour is precious. It is now fourteen days since they were set on foot. Sure they cannot be much longer in preparing."

The $240 were sent by messenger.

Washington, writing to his brother, John Augustine, October 13, 1775, said:

"Finding that we were in no danger of a visit from our neighbours, I have fitted out and am fitting out several privateers with soldiers who have been bred to the sea; and I have no doubt of making captures of several of their transports, some of which have already fallen into our hands laden with provisions." [*Am. Ar.*, 4th Series, Vol. III, p. 1055.]

The armed vessels fitting out were the schooners "Lynch,"

Captain Nicholas Broughton, and the "Franklin," Captain John Selman. They were named after two of the Committee of Congress consulting with Washington. Later the "Harrison," after the third member of the Committee, was fitted out as were the "Lee," the "Warren" and the "Washington." About January 1, 1776, the "Hancock" was added. Of this fleet Washington had the entire command, Moylan conducting the correspondence after having fitted the vessels out. In January Captain Manly was appointed by Washington commodore of the fleet.

It is to be remembered these were not Continental vessels but "privateers," as Washington called the vessels which he fitted out to prey upon unarmed supply vessels. Congress was at this time promoting the fitting out of armed vessels under Continental authority to attack armed vessels of the British. The first was the "Lexington," to which Captain John Barry was appointed, on December 7, 1775, Captain, and the "Reprisal," of which Captain Wickes was made Commander the same day.

On October 16, 1775, Washington sent orders to Moylan and Glover, then at Marblehead, that the two vessels must be immediately dispatched. At the same time he ordered Captain Broughton to "proceed to intercept two north-country brigantines of no force" bound for Quebec laden with 6,000 stand of arms, powder and other stores. Captain Selman received the same order but to act with Broughton whom he was to consider as Commodore.

But the vessels were not ready and would not be for two weeks. "If not soon at sea," wrote Colonel Reed, "we shall heartily regret it was ever undertaken." On the 19th he wrote Moylan: "For God's sake hurry off the vessels; transports without convoy arrive every day at Boston." Moylan and Glover that day wrote Reed to inform Washington that the two vessels "will both be ready to sail to-morrow. Mr. Moylan has the pleasure to inform His Excellency that the flour is all arrived."

MOYLAN AND A FLAG.

On the 20th Reed wrote to Moylan and Glover that British squadron is bombarding Falmouth and Portsmouth. Our vessels must be careful how they fall in with them. "Please fix upon some particular color for a flag and a signal by which our vessels may know one another. What do you think of a flag with a white ground, a tree in the middle, the motto 'Appeal to Heaven'? This is the flag of our floating batteries. We are fitting out two vessels at Plymouth and when I hear from you on this subject I will let them know the flag and signal, that we may distinguish our friends from our foes."

Next day from Beverly Moylan and Glover notified Reed: "The schooners sailed this morning. As they had none but their old colours, we appointed them a signal, that they may know each other by, and be known by their friends—the ensign up to the main topping lift."

"Mr. Moylan has ordered 300 bbls. of flour that was at Ipswich to this place for the use of the Navy; it saves some miles of land carriage." Among the articles wanted "immediately" were "two signal flags."

"THE SPIRIT OF EQUALITY."

The vessels of Broughton and Selman having been despatched two others were being fitted out. Moylan, on the 24th, wrote Reed:

"I wish with all my soul that these two vessels were dispatchd cheifly for the publick service, & also that I may have the pleasure of seeing my friends Mr. Lynch & Col. Harrison, I want much to be introduced to Doctor Franklin for whom I have many years a vast veneration. I think they will be off on Thursday evening, if they are, I will be with you on Friday.

"Col. Glover shewed me a Letter of yours which has mortified him much, I realy & sincerely believe he has the cause much at heart, & that he has done his best (in the fitting out these four last vessells), for the publick service you cannot conceive the difficulty the trouble & the delay there is in procuring the thousand things necessary for one of these vessells, I dare say one of them might be fitted in Philadelphia or New York in three days, because you would know where to apply for the different articles but here you must search all over Salem, Marblehead, Danvers & Beverly for every little thing that is wanted. I must add to these the jobbing of the carpenters, who are to be sure the idlest scoundrels in nature, if I could have procured others, I should have dismissed the whole gang of them last Friday, & such religious rascalls are they, that we could not prevail on them to work on the Sabbath. I have stuck very close to them since & what by scolding & crying shame for their torylike disposition in retarding the work I think they mend something.

"There is one reason, & I think a substantial one, why a person born in the same town or neighborhood should not be employed on publick affair of this nature in that town or neighborhood, it is that the spirit of equality which reigns thro this country, will make him afraid of exerting that authority necessary for the expediteing his business, he must shake every man by the hand, & desire, beg & pray, do brother, do my friend, do such a thing, whereas a few hearty damns from a person who did not care a damn for them would have a much better effect, this I know by experience, for your future government—indeed I could give other reasons, but I think this sufficient."

Concerning this spirit of "equality" which pervaded the army General Wilkinson in his *Memoirs* states: "On entering the camp near Boston, I was struck with the familiarity which prevailed among the soldiers and officers of all ranks; from the Colonel to the private, I observed but little distinction; and I could not refrain from remarking that the military discipline of their troops was not so conspicuous as the civil subordination of the country in which I lived."

Washington, writing to Congress, September 21, 1775, relative to the pay of officers, declared it "is one great source of that familiarity between officers and men which is so incompatible with subordination and discipline."

On October 25, 1775, Colonel Reed wrote Glover and Moylan at Salem or Marblehead: That intelligence from Boston was that "a transport with 1200 bbls. of powder, without convoy or force, had been missing and expect to fall in our hands." A large schooner carrying ten guns would be fitted out. "I have given them the signals." [*Washington Papers,* VIII.]

On 27th Moylan, from Beverly, wrote Reed that young Captain Glover had returned without the 300 swivel shot—the "most material article." He says there were none, but there were plenty of "4-oz. bullets which, if he had had one ounce of sense must have known would answer all the purposes."

"Captain Manly's vessel is all ready. We now only wait the collecting together his hopeful crew to send him off. I have declared that if there are even thirty on board to-morrow morning and the wind proves fair that he shall hoist sail. I am much grieved that I had not the pleasure of seeing Mr. Lynch and Col. Harrison. I regard them highly. Dr. Franklin is going and you are also on the wing. Every one engaged in this contest must sacrifice their private satisfaction to the public good." [*Washington's Papers,* Vol. VIII.]

On October 28th Moylan and Glover wrote Colonel Reed that Captain Glover had brought "all the things we wrote for" except the 300 swivel shot which were not to be had. There was a shortage of ammunition for Captain Adams whose vessel was ready but there was no appearance of him or his men. "Captain Manly is off and only waits a fair wind to proceed to sea."

Reed replied on the 30th and added, "I am just starting for Philadelphia."

Washington felt the loss of the services of Reed very much. Edmund Randolph and George Baylor had, on August 15th, been appointed to aid Reed. After Reed's retirement Robert Hanson Harrison, on November 5th, became Secretary. At this time Moylan was at the Camp at Cambridge as Mustermaster-General, but frequently acting as Secretary *pro tem* for the General.

On November 20, 1775, Washington wrote Reed:

MOYLAN "VERY OBLIGING."

"You can judge that I feel the want of you when I inform you that the peculiar situation of Mr. Randolph's affairs obliged him to leave this place soon after you did; that Mr. Baylor, contrary to my expectations, is not in the smallest degree a penman, though spirited and willing; and that Mr. Harrison, though sensible, clear and perfectly confidential, has never yet moved upon so large a scale as to comprehend at one view the diversity of matter which comes before me, so as to afford that ready assistance, which every man in my situation must stand more or less in need of. Mr. Moylan, it is true, is very obliging; he gives me what assistance he can; but other business must necessarily deprive me of his aid in a very short time." [*Amer. Arch.,* 4th Series, Vol. III, p. 1619.]

To show the assistance of Moylan at this time observe the numerous letters written by him on Washington's affairs which appear in this narration.

WASHINGTON LAMENTS THE DEARTH OF PUBLIC SPIRIT.

Washington wrote to Reed on 28th November, 1775:

"Such a dearth of public spirit and such want of virtue, such stock jobbing and fertility in all the low arts to obtain advantages, of one kind or another, I never saw before, and pray God I may never be witness to again. . . . Could I have forseen what I have

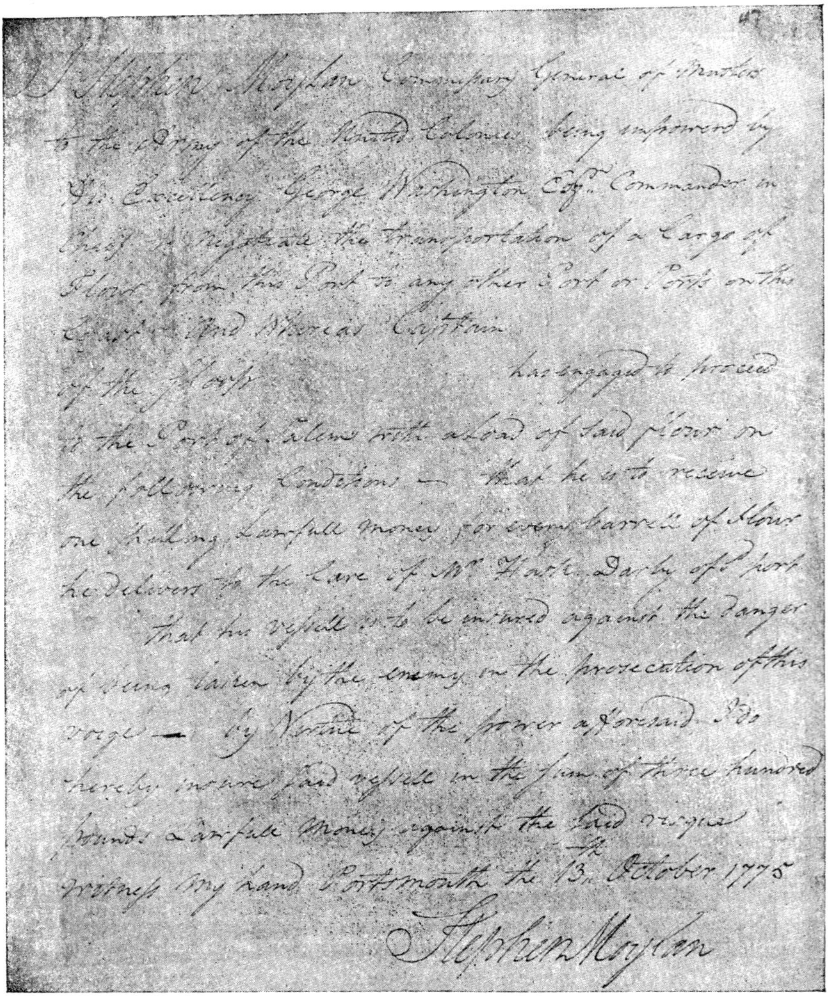

I Stephen Moylan Commissary General of Musters
to the Army of the United Colonies being impowered by
His Excellency George Washington Esqr. Commander in
Chief to facilitate the transportation of a Cargo of
Flour from this Port to any other Port or Ports on the
Coast And Whereas Captain
of the Sloop has engaged to proceed
to the Port of Salem with a load of said flour on
the following Conditions — that he is to receive
one shilling Lawfull money for every barrell of flour
he delivers to the Care of Mr. [illegible] Darby of sd port
That the vessell is to be insured against the danger
of being taken by the enemy in the prosecution of this
voyge — by Virtue of the power aforesaid I do
hereby insure said vessell in the sum of three hundred
pounds Lawfull Money against the said voyage
Witness my hand Portsmouth the 13th October 1775
Stephen Moylan

HANDWRITING OF MOYLAN

and am like to experience no consideration upon earth should have induced me to accept this command. A Regiment or any subordinate department would have been accompanied with ten times the satisfaction—perhaps the honour. . . . I find it necessary that the aids to the commander-in-chief should be ready at their pen to give that ready assistance expected of them. What can your brethren of the law mean by saying your perquisites as Secretary must be considerable? I am sure they have not amounted to one farthing." [*Life Reed,* I, 132; *Sparks',* III, 178.]

MOYLAN'S LETTERS.

From Cambridge, November 4, 1775, Moylan wrote the Committee of Safety of Dedham, Mass., that Washington had received a letter from David Parker in which he most pathetically deplores his situation and made "the most solemn assurances of contrition for the part he acted and strong declarations of his regard for the liberties of his Country, he prays that the arrest under which he now is may be removed."

Washington had "no objection to his enlargement," provided he "can make it clear to the Committee that he is no longer inimical to the Country." Moylan likewise wrote Parker to the same purport. He also expressed to Samuel Goodwin, Washington's pleasure at his having supplied General Arnold with the plans of his route to Quebec—that "if it should be found necessary to lay out the road," Washington "won't be unmindful" of Goodwin's services.

On November 5, 1775, Moylan wrote Bartlett relative to the capture of a sloop from Boston by Captain Brown. That Washington directed an inventory be made of the goods. That the General would make such satisfaction to the two resolute fellows who first ventured on board as is in such cases proper and customary. "Providence," concluded Moylan, "has sent us a good supply of wine by a vessel from Philadelphia, being stranded at Eastham with 120 pipes bound to Boston of which 118 are ordered to Cambridge." [*Am. Ar.*, 4th Series, Vol. III, p. 1367.]

Later he wrote Watson, Agent at Plymouth, that the wine belonged to Thomas Satler of Philadelphia and was not intended for the enemy. So he was to have it stored and await directions.

The following day he notified Watson to send the wine to Washington's camp, where it would "be sold for the public use and bring a good price." That the General, on account of the advanced season and the difficulty of procuring cannon, would order out no more armed vessels. The intention in fitting out these vessels is not to attack the armed but to take unarmed vessels. He wished Captain Coit success.

SUCCESS OF THE PRIVATEERS.

Capain Coit did have a success. That day he, after a cruise of thirty-six hours, brought into Plymouth the schooner "Industry" and the sloop "Polly," both from Nova Scotia bound for Boston with cattle and provisions for the garrison. [*Am. Ar.*, 4, III, 1328.]

Watson, the Agent at Plymouth, pastured the cattle there and sent the prisoners to Washington at Cambridge. On November 10th Moylan notified the General Court of Massachusetts of the capture, sending, by "command of Washington, the papers in the case" and also Jabez Hatch "who appears to be a noted Tory and is owner of one of these vessels," whom Moylan requested the Council to do with as seemed proper. [*Ibid.*]

On November 8th Moylan wrote Watson that Washington desired him to "sell the articles found on the two prizes." "We shall soon hear of Captain Manly's being successful and that Captain Coit is again at sea, to pursue his good fortune. The men and ammunition for the 'Washington' will set out to-day. I recommend all possible dispatch to Captain Martindale."

The same day he wrote John Brown at Providence, R. I., that the General had ordered Colonel Gridley to "procure the cannon necessary for the use of the camp."

Moylan also, on the same day, wrote Captain E. Bowen, Jr., "time is very short for expecting more prizes, the season being so far advanced. This is one reason his Excellency's determination not to fit out more cruisers for the present."

CAPTURES WITHOUT AUTHORITY.

On November 4, 1775, a sloop, the "North Briton," was captured by "two resolute people in a small boat at one of the islands called Misery."

On November 8th, off Beverly, a schooner laden with supplies was captured by "fifteen men."

The Commander refused to show the ship's papers to Bartlett, the Prize Agent, who wrote to Washington for directions, "for if I have no power to make such demand I make myself ridiculous in the eyes of the world." On November 11th Moylan wrote that Washington's advice was to "have nothing to do with such vessels by any authority under him. Don't trouble yourself or the General with a litigious dispute. In short, get rid of the trouble in the best way you can and let us hear nothing further thereon."

The captures, it is to be noted, were made by individuals not having authority to make captures. Yet on 15th Moylan wrote Bartlett that as the goods on the schooner must be sold at vendue, he requested that all the claret be purchased for him. "As it is a liquor not much used in this country, it will probably sell cheap. Should you have occasion for part of it yourself, you will, by all means, keep whatever you may want." [*Mag. Am. His.*, May, 1890.]

On November 9th Moylan wrote Captain Jonathan Glover at Marblehead that the General directed that the persons belonging to the two schooners sent into Marblehead be discharged, but that bondsmen be had that they will not leave the district nor give information concerning the destination of Captains Broughton and Selman; they were to be well looked after. The wood on the sloop brought in by Manly was to be sold and the vessel laid up until it was determined whether she was a prize or only a recaptured vessel.

The next day, 10th, Moylan wrote Wentworth, Chairman of the Committee at Newburyport, that the Penobscot Indians needed powder; that "if not supplied by us with some, they will make application to the enemy, who, no doubt, would gladly embrace such an opportunity of making them friends." Washington directed that two barrels be given them out of the stock of the Committee, which "he would replace if it is not done by the legislature."

On November 16th Moylan wrote Watson at Plymouth that Washington ordered that those captured by Captain Coit be given their bedding and wearing apparel, but he was at a loss to know whether it is customary to return the money found on prisoners; these people do not merit any indulgence. "His Excellency would rather err on the side of mercy than that of a strict justice. Let me know your opinion in this matter."

Watson, on November 23d, replied advising that the matter be referred to the Committee of Safety.

On October 18, 1775, Falmouth was destroyed by the British; they burned five hundred houses, fourteen vessels, and all without loss to themselves. "A full demonstration that there is not the least remains of virtue, wisdom or humanity in the British Court."

Measures were taken after this devastation to defend the town against further attacks. On November 24th Moylan wrote Samuel Freeman of Falmouth that Washington approved of what had been done in defence of the port, adding: " It is incumbent on the people of the country to exert themselves for their and the publick defence. The Congress are so much of that opinion that they have recommended it to each of the Colonies to provide for their particular internal safety." [*Am. Ar.*, 4, III, 1666.]

POPISH BIGOTRY.

While Moylan, a Catholic, was so active, Jacob Bayley was writing from Newbury to Colonel Little relative to Canadian affairs, saying: " Our people, doubtless, are amongst them which will wear out their Popish bigotry; until that is done no great trust to the French." [*Ibid*, 1664.]

Instead of wearing out the " bigotry" of the Canadians, that kind of language, and conduct to accord with it, wore out the friendship and aid of the Canadians.

General Howe at same time wrote to the Earl of Dartmouth: " This army, though complete in the Spring, must have six or seven thousand recruits and chiefly of the worst kind if chiefly composed of Irish Roman Catholics, certain to desert if put to hard work, and, from their ignorance of arms, not entitled to the smallest confidence as soldiers." [*Am. Ar.*, 4, III, 1673.]

It was not very long before General Howe himself, when in Philadelphia, undertook to form a regiment of such Irish Roman Catholics as could be recruited or induced to desert from Washington's Army at Valley Forge. He had confidence enough in such as he could get to avail himself of their services.

Captain Adams had taken a vessel with a cargo of potatoes and turnips, which seems not to have been regarded as a valuable prize, in contrast with one taken by Captain Manly. Moylan, writing from Washington's camp on December 1, 1775, to Colonel Joshua Wentworth, at Portsmouth, saying:

" We are all flushed with the agreeable account of Captain Manly's having taken a prize of the utmost consequence, which made us look over the potatoes and turnips of Captain Adams, but, now being a little cool, I assure you I do not think Adams' *bon fortune* so despicable. Though of little value to us, it is depriving

218978

the enemy of what to them would be of consequence. As to the prisoners, I wish you had kept or discharged them. . . . You will please dispose of the cargo by the "Rainbow" and lay up the schooner until further orders. Though, if a reasonable price can be got for her, you may dispose of her. I hope Adams will soon take such a prize as Manly has. I really believe the cargo could not cut short of £10,000 sterling. To us it is invaluable."

On December 1st Moylan wrote Wm. Watson, Plymouth:

"Capt. Coit's Lieutenant has been here and gives an account of his schooner being so old and crazy as to be unfit for the service he is employed in. If there was a possibility of fitting a better vessel out in six or eight days and removing the guns from on board the schooner, His Excellency, would be glad it could be done, as there are store ships and transports expected all this month."

Relative to Captain Martindale's finding it impossible to get men, Moylan declared "that the deficiency of public spirit in this country is much more than I could possibly have an idea of. Manly's crew will make their fortune by his activity—a quality Martindale is deficient in—get out his brigantine let the expense be what it will." [*Am. Ar.*, 4, IV, 153.]

On December 2d Moylan wrote Bartlett and Glover, Agents of the brigantine "Nancy" at Beverly or Gloucester, that as it was "under consideration of Congress, the mode of disposing of such vessels and crews as are taken supplying the enemy it is his Excellency's pleasure that Captain Hunter and his crew return to you, that their private adventure be given them with liberty to dispose thereof as they think proper, that they be treated with all humanity, due to fellow-citizens in distress."

On December 4th Moylan sent the Massachusetts Council the names of the prisoners taken on the sloop "Polly" and the sloop "Success" and also five sailors of the "Canceaux" man-of-war. One was named Pat Burns.

On December 4th Moylan, by Washington's "command," notified Bartlett that concerning the capture of the sloop "Concord" from Glasgow with goods for Boston that though the enemy were daily seizing our vessels and that reprisals ought to be made, he did not think he had authority to declare lawful captures could be made of vessels transporting goods from English or British owners for their agents here, and had referred the matter to Congress.

"OLD PUT."

When Manly's captured stores were taken to Washington's camp at Cambridge the scene is thus described by Moylan in letter of December 5th to Colonel Reed:

"I would have given a good deal that you was here last Saturday when the stores arrived at camp. Such universal joy ran through the whole as if each grasped victory in his hand. to crown the glorious scene there intervened one truly ludicrous, which was old Put mounted on a large mortar which was fixed in its bed for the occasion, with a bottle of rum in his hand, standing parson to christen, while godfather Mifflin gave it the name of Congress. The huzzas on the occasion I dare say were heard through all the territories of our most gracious sovereign in this Province.

"The time of the Connecticut troops' enlistment being expired, the scoundrels are deserting the lines before we are prepared for such a defection." [*Reed's Life,* I, 134.]

General Howe, in reporting to the Earl of Dartmouth, December 3d, the capture of the "Nancy," stated she had "4000 stand of arms complete, 100,000 flints, a 13-inch mortar with other stores in proportion." "The capture" is rather unfortunate to us. "The Rebels" are now furnished with all the requisite for setting the town on fire.

These supplies were those taken on the "Nancy" captured by Manly. This capture was to Washington "an instance of divine favour—for nothing surely ever came more *apropos.*" Though Manly "unluckily missed the greatest prize in the world—their whole ordnance—the ship containing it being just ahead—but he could not have got both." [Reed's *Life,* I, 132.]

The "Nancy" sailed under convoy of the "Phœnix," man-of-war, and on November 27, 1775, General Howe wrote Earl of Dartmouth that she was "the only ordnance store ship missing. . . . The Rebel's cruisers are watchful and have already been too successful, and will probably do much more mischief unless the King's ship can contrive to cut them off." [*Am. Ar.,* 4, III, 1679.]

Next day, 5th, he informed the Committee of Safety of Salem that the brig "Kingston Packet" had been taken at Barington, Nova Scotia, on suspicion of being engaged in business contrary to the Association of the United Colonies. Washington directed the Committee to determine the case.

Again, on December 5, 1775, he notified Captain Peleg Wadsworth that Washington desired he would examine the harbour of Cape Cod and see what fortifications may be necessary for its defence and to report thereof.

On 6th to Salem Committee that Washington would contribute to the safety of the town when he can do it, consistent with that attention he must pay to the defence of the whole.

Same day to Bartlett sending $2000 for military stores.

On 8th to the Salem Committee delivering up the cargo of a vessel.

Same day to Glover at Marblehead, that the affair of the "Kingston Packet" be managed "so that Headquarters may hear no more of it."

December 9th—To Massachusetts Council sending four pilots taken by Captain Coit on board vessels taking supplies to the enemy in Boston.

TWO GENTLEMEN FROM ANTIGUA.

On December 10th Moylan wrote Bartlett regarding the Captain of a captured vessel, that it was "very unlucky the Captain threw his papers overboard—and if it were true that this was done after he was made a prize of, he deserves to be severely punished and in any other war he would suffer death for such an action, but we must show him and all such who fall into our hands that Americans are humane as well as brave. You will, therefore, treat the prisoners with all possible tenderness."

There were on board two gentlemen from Antigua—Mr. Burke and Mr. Gregory. Mr. Burke was "strongly recommended to our good friends in Boston, although not friendly to American liberty, he still has a character as a gentleman," and it was Washington's orders that he be treated as such.

Mr. Gregory was "going to serve on a man-of-war. Both were to be paroled. . . . There are limes, lemons and oranges on board, which being perishable you must sell immediately. The General will want some of each as well as the sweetmeats and pickles, as his lady will be here to-day or to-morrow. You will please pick up such things as you think will be acceptable to her He does not mean to receive anything without payment."

Washington ordered that prizes be not visited by people from the shore or from the armed schooners, "that embezzlement be particularly guarded against." [*Mag. Am. His.*, 1890, p. 414.]

The same day he wrote Bartlett by Captain Adams who was anxious to go on a cruise, "give him every assistance in your power—indulge him and let him proceed to sea."

December 13, 1775, Moylan wrote to Watson at Plymouth: "Captain Manly's good fortune seems to stick to him; he has taken three more valuable prizes. This shows what advantage these vessels would be, if the commanders were all as attentive to their duty and interest as Manly is. There runs a report that one of our little fleet is taken and carried into Boston. We shall be uneasy till we hear from Martindale, as he is the one suspected."

December 14th, wrote Colonel Wentworth at Portsmouth: "Captain Manly has been very successful, having taken three more prizes. I wish the commanders of the rest of our little fleet were as active; if they were, we could conquer our enemies without loss of blood."

THE NEW ARMY AND THE NEW FLAG.

On January 1, 1776. This day the newly recruited army—the really first Continental Army—was paraded by Washington and the new Union flag hoisted on Prospect Hill in compliment to the United Colonies. This Union flag was the thirteen stripes with the crosses of England and Scotland in the canton.

Though England's cross was displayed, the day really marks the separation of America from England in the mind of Washington, Moylan and many others. On that day the King's speech "full of rancour and resentment and declaring that vigorous measures would hereafter be pursued to crush the foul and unnatural rebellion and giving, as Washington declared, "the ultimation of British justice." was by General Gage sent out of Boston so as to be distributed among Washington's men. When the Union flag was displayed by Washington it was taken by the British "as a token of the deep impression the King's speech had made on us," wrote Washington, "and as a signal of submission." [*Am. Ar.,* 4, IV, 570.]

Instead of submission, "the Speech but strengthened the Independency thought which had been growing in many minds.

FOR INDEPENDENCE.

Colonel Moylan was eager for Independence six months before it was Resolved upon and the Declaration made. Writing to Colonel Reed from Cambridge we may be sure he expressed no opinion on Independence adverse to those held by Washington. He said, January 2, 1776:

"The Congress is still sitting and I am glad of it. Will they now hesitate? Look at the King's speech. Will they not immediately send embassies to some foreign powers? Will they not declare what his Most Gracious Majesty insist on they have already done? Will they not strain every nerve to accomplish it? Are there remaining any hopes of a desirable alternative? They are men of sense and will act right.

"I should like vastly to go with full and ample powers from the United States of America to Spain, if my old friend Wall is still living, and he had influence, I am sure I could do service there. [*Reed's Life,* p. 139.]

Who was this "old friend Wall"?

"We really are tired of inaction. . . . Why are not vessels sent out this winter from those ports which will continue open by God Almighty's permission? Will not Congress follow the good example of the Almighty and open them all to the world? The King's speech is the key to open all ports.

"All the vessels are now in port—the officers and men quitted them! What a pity, as vessels are every day arriving—the chance of taking any is pretty well over, as a man-of-war is stationed so as to command the entrance of Beverly, Salem and Marblehead. We must have ships to cope with them. I shall try and get some of them to sea while the weather continues mild. Five hundred men of the Irish reinforcement arrived within these few days at Boston."

Again he returned to the subject of Independence when, on January 30, 1776, he wrote Colonel Reed:

"BOLDLY DECLARE INDEPENDENCE."

"Shall we never leave off debating and *boldly declare Independence.* That and that only will make us act with spirit and vigour. The bulk of the people will not be against it—but the few and timid always will,—but what can be expected of a contrary conduct? Can it be supposed possible that a reconciliation will take place after the loss of blood, cities and treasure already suffered, but the war must come to every man's home before he will think of his neighbour's losses." [*Life of Reed,* I, 160.]

Robert Morris, writing from the Falls of Schuylkill, July 20, 1776, to Colonel Reed, said: "Remember me to Colonel Moylan."

We have seen Moylan busy while away from the camp at Cambridge in fitting out vessels, and have noted his industry while at camp in assisting Washington in his correspondence. At times Washington was, as he wrote to his absent Secretary, Reed, January 23, 1776, "so much taken up at his desk that I am obliged to neglect many other essential parts of my duty," as "Mr. Moylan's time must now be solely employed in his department of Commissary." He had hinted to Moylan and to Mr. Harrison "that as they really had a great deal of trouble each of them should receive one-third of" Reed's "pay, reserving to yourself the other third, contrary to your desire."

On that date both Harrison and Moylan were not at camp; the former had written "to ascertain if his return cannot be dispensed with," and Moylan was away endeavoring to hasten the sailing of the fleet to attack British unarmed supply vessels. On 19th he reported to Washington who received it on 25th, "but his time was so employed in despatching expresses to sundry places" that he could not send reply until the next day, when Harrison wrote Moylan:

"His Excelency is much pleased that our Fleet is likely to get out again, & wishes your Return as soon as you have dispatched them. Herewith you will receive the Commission you wrote for; also a Copy of private Signals used by the King's ships in the American Service, which his Excellency desires you to furnish each of our Captains with & to return the Copy sent, when you come back.

"As to the Attempts on the Fowey, he thinks, that the situation of our Affairs at present will not justify it. But on your Arrival, will consider more of it, should there be the same Prospect of success that you apprehend there is now.

"The Report you have had is too true; but not so bad, we hope, as you have heard. However, it is certain that the great & gallant Montgomery, is no more. He, with his Aide-de-Camp, Captain McPherson and Captain Cheeseman of New York, fell the first fire; also Colonel Arnold is wounded in the Leg."

Moylan returned to camp on January 24, 1776, and "by command from his Excelency" he sent Major Hawley of Berkshire a commission "for such person as he may think qualified to muster the regiment raising in Hampshire and Berkshire."

On 26th Moylan wrote the General Court of Massachusetts that Washington desired to know the mode of drawing the money the Province had offered to advance him for the use of the United Colonies.

CAPTURE OF THE "HOPE" FROM IRELAND.

In August, 1775, the ship "Hope" from Cork, Ireland, came up the Delaware River, with Major Christopher French of the 22d Regiment, British Army; Ensign John Rotton, 47th Regiment, and Terence McDermott [in *Washington's Papers* called Cadet William McDermott] and two privates. The "Hope" was seized by one of the Pennsylvania armed boats and the officers and men made prisoners. They had come "hither without any knowledge of hostilities." They were, on August 12th, brought before the Pennsylvania Committee of Safety and after examination it was found they had come "hither with an intention of joining the Ministerial Army at Boston, under the command of General Gage, who is now acting in a hostile and cruel manner against his Majesty's American subjects." The Committee paroled the officers but held their goods, both subject to Washington's orders. [*Am. Ar.*, 4, III, 499.]

On August 25th Washington directed the prisoners be sent to Hartford, Connecticut. They were so sent. Washington directly, and also through his Secretary, Reed, had much correspondence with Major French concerning his application to be allowed to wear his sword, to be exchanged, or to be permitted to go to Ireland. Details of this can be had in the Connecticut Historical Society's *Collections*, Vol. I, and in the *Washington's Papers* in the Library of Congress. Our present concern in the case relates to the connection of Stephen Moylan therewith as Secretary to Washington.

MOYLAN "TUTORS" MAJOR FRENCH.

On February 10, 1776, Moylan, by direction of Washington, wrote Major French:

"SIR:—Your repeated letters to Gen. Gates desiring liberty to go to Ireland on your parole were laid before his Excellency. I have it in command from him to inform you that he does not think himself authorized to grant license to any one to depart this Continent—that power is lodged only in the hands of the Congress.

I am also commanded to tell you that the General is surprised a gentleman of Major French's good sense and knowledge should make such a request. Let him compare his situation with that of such gentlemen of ours who by the fortune of war have fallen into the hands of their enemy. What has been their treatment? Thrown into a loathsome prison and afterward sent in irons to England. I repeat—let the Major compare his treatment with theirs and then say whether he has cause to repine at his fate."

French replied that General Gates had written him that Washington was ever willing to grant indulgences to gentlemen officers, "but at present could not comply." He added: "Mr. Moylan was pleased to tutor me with a parallel upon the different treatment of prisoners which apears to me to have been lugged in (like the tailor in Lethe) by head and shoulders, as it was entirely foreign to the subject, since I did not complain of bad treatment. Why does Mr. Moylan, whom I don't know, write me upon a topic which I write to General Gates about and why, at least, does he not assign a reason for General Gates not writing or signing the letter." [*Con. His. Soc. Col.,* I, 212.]

On November 15th Major French escaped, leaving behind him a journal.

APPOINTED SECRETARY AND AIDE TO WASHINGTON.

Though Moylan held the rank of Muster-Master he had been attending to the duties of that position and also frequently acting as Secretary to the Commander-in-Chief, he was not, officially, Secretary until March 5, 1776, and the next day he was named as an Aide. [Ford's *Washington,* XIV.] Colonel Knox "being desirous of it," Washington wrote Reed on March 7th.

An Aide ranked as Lieutenant-Colonel.

On March 7th Washington wrote Colonel Joseph Reed: "I have appointed Mr. Moylan and Mr. Palfrey my Aides-de-Camp so that I shall, if you come, have a good many writers about me."

Moylan was Secretary to Washington when, on St. Patrick's Day, 1776, the British evacuated Boston. As Secretary he received the appeal sent out of Boston to Washington on March 8th by four leading citizens, stating:

"As his Exc'y Gen. Howe is determined to leave the Town . . . a number of the respectable Inhabitants (being very anxious for its preservation & Safety) have applied to Gen. Robertson," who . . . has communicated the same to Gen'l Howe who has assured him, "That he has no intention of destroying the town" "unless the Troops under his Command are molested during" "their Embarkation or at their Departure by the Armed Forces" "Without"——

If such an Opposition should take Place we have the greatest Reason to expect the Town will be exposed to entire Destruction; And . . . we beg we may have some Assurances that so dreadful a Calamity may not be brought on, by any Measures Without."

"This paper," wrote Washington to Reed, "seems so much under covert, unauthenticated and addressed to nobody, that I could take no notice of it; but shall go on with my preparations as intended." [*Sparks,* III, 311.]

Washington had made it known that he was determined to have possession of the City even if he had to burn it. Hence the alarm among its inhabitants and the consequent more hasty evacuation.

SCARCITY OF POWDER.

But for the scarcity of powder it is probable that Washington would have long before the evacuation have bombarded the City. On January 2, 1776, Moylan wrote Colonel Reed: "It will be possible to bombard Boston; give us powder and authority, for that you know we want as well as the other. Give us these and Boston can be set in flames." [*Reed's Life,* I, 137.]

This scarcity of powder existed all the time Washington was besieging Boston.

On August 4, 1775, Washington informed Congress he had not "more than nine rounds of powder a man" and that "our situation in the article of powder is much more alarming than I had any idea of." [*Am. Ar.,* 4, III, 28.]

The same day he wrote Governors Cooke and Trumbull: "Our necessities in the article of powder and lead are great—the case calls loudly for the most strenuous exertions of every friend of his country and does not admit of the least delay." He approved of an endeavor to capture powder at Bermuda where there was a considerable magazine. "No quantity however small is beneath our notice." [*Ibid,* 36-38.]

NOT THE LIKE OF IT IN HISTORY.

On January 4, 1776, he wrote Congress:

"It is not in the pages of history, perhaps, to furnish a case like ours; to maintain a post within musket shot of the enemy, for six months together, without powder, and, at the same time, to disband one army and recruit another, within that distance of twenty-odd British regiments, is more, probably, than ever was attempted. But, if we succeed as well in the last, as we have, heretofore, in the first, I shall think it the most fortunate event of my whole life." [*Am. Ar.*, 4, IV, 567.]

The same day he wrote his former Secretary, Joseph Reed, at Philadelphia:

"Search the volumes of history through, and I very much question whether a case similar to ours is to be found; namely, to maintain a post, against the flower of the British troops, for six months together, without powder, and then to have one army disband and another to be raised, within the same distance of a reinforced enemy. . . . For more than two months I have scarcely emerged from one difficulty, before I have been plunged into another. How it will end, God, in His goodness will direct. I am thankful for his protection to this time."

POWDER! POWDER!

The *Journal of Elias Boudinot* relates:

"When our army lay before Boston in 1775 our powder was so nearly expended that General Washington told me that he had not more than eight rounds a man altho he had then near fourteen miles of line to guard and that he dare not fire an evening or morning gun. In this situation one of the Committee of Safety for Massachusetts, who was privy to the whole secret, deserted and went over to General Gage, and discovered our poverty to him. The fact was so incredible that General Gage treated it as a stratagem of war and the informant as a spy; or coming with the express purpose of deceiving him and drawing his army into a snare, by which means we were saved from having our quarters beaten up."

"If we had powder," wrote Moylan to Reed, on January 30, 1776, "I do believe Boston would fall into our hands."

"OLD PUT" AND POWDER.

On February 1, 1776, he wrote from Roxbury: "The bay is open, everything thaws except old Put. He is still as hard as ever.

crying out, 'Powder! Powder! Ye gods, give me powder.'" [*Potter's Mo.,* VI.]

As Putnam was born in 1718 he was nearly seventy years "old," but yet young in enthusiasm and zeal, as his cry for "Powder! Powder!" testifies.

Washington, on 31st March, 1776, wrote his brother, John Augustine: "I have been months together with (what will scarcely be believed) not thirty rounds of musket cartridges to a man; and have been obliged to submit to all the insults of the enemy's cannon for want of powder, keeping what little we had for pistol distance. We have maintained our ground against the enemy, under this want of powder, and we have disbanded one army and recruited another within musket shot of two and twenty regiments, the flower of the British army, whilst our force has been but little, if any, superior to theirs; and, at last, have beaten them into a shameful and precipitate retreat out of a place the strongest by nature on this Continent, and strengthened and fortified at an enormous expense." [*Sparks,* III, 340.]

"I believe I may with great truth affirm that no man, perhaps, since the first institution of armies ever commanded one under more difficult circumstances than I have done." [*Ibid,* 343.]

Washington, in April, 1776, moved his army from Boston to New York to counteract the movements of the enemy designed to make the Hudson River the campaigning region, so as to cut the Colonies in two, as it were, by separating New England from the other Colonies.

General Charles Lee, second in command of the American army, then at Williamsburg, Virginia, on May 10, 1776, wrote General Washington, then at New York: "I am well pleased with your officers in general and the men are good—some Irish rascals excepted." He closed by sending "My love to Moylan." [*Letters to W.,* I, 203.]

APPOINTED QUARTERMASTER-GENERAL.

Colonel Moylan was, at this time, Muster-Master General, an Aide to Washington and also his Secretary. On June 5th he was, by Congress, elected Quartermaster-General to succeed Colonel Thomas Mifflin.

Congress also *Resolved,* That Stephen Moylan, Esq., have the pay of eighty dollars a month and the rank of Colonel.

On June 10th Congress received "a letter from Stephen Moylan expressing his grateful thanks to Congress for appointing him to the office of Quartermaster-General."

Moylan was succeeded as Secretary to Washington on May 16th by Robert Hanson Harrison who had been also acting as Secretary from November, 1775.

Of Moylan and Harrison Washington wrote: "Mr. Harrison is the only gentleman of my family that can afford me the least assistance in writing. He and Mr. Moylan have hitherto afforded me their aid and they really had a great deal of trouble."

But once again Moylan acted as Secretary to Washington when, on June 13, 1777, Washington, at Morristown, New Jersey, wrote General Howe, the British General, commanding at New York, relative to the Exchange of General Charles Lee and the treatment of prisoners. It was sent by Washington to Congress on 14th. [*Ford,* V, 168.]

On the return of Charles Carroll of Carrollton, as one of the Commissioners to Canada to secure the aid or neutrality of the Canadians he, at New York, on June 9, 1776, "Waited on General Washington; saw Generals Gates and Putnam and my old acquaintance and friend Mr. Moylan."

After arrival at Philadelphia to report to Congress he wrote, on June 14th, to General Gates to present his "respectful compliments to General Washington and remembrances to General Mifflin and my friend Moylan." [Rowland's *Carroll,* I, 176.]

On June 10, 1776, Congress resolved that the Quartermaster-General be directed to procure and forward such tents, clothing and untensils as are wanted for the army in Canada, subject to the directions of the Commander-in-Chief.

On June 17, 1776, Congress Resolved: That Mr. James Mease be directed to purchase and forward to the Quartermaster-General in New York as much cloth for tents as he can procure.

On June 24th, information having been given Congress that a quantity of tents which were sent from Philadelphia for the use of the Colony of Massachusetts Bay have been stopped at New York by order of the Quartermaster-General, it was Resolved that the President write to the General on this subject and desire him to order the said tents to be re-delivered and forwarded to the Colony of Massachusetts Bay immediately.

On June 29th Washington issued Order: "The General expects that all soldiers who are entrusted with the defence of army work will behave with great coolness and bravery and will be particularly careful not to throw away their fire."

Washington expecting a combined attack by General Howe with ten thousand men and Admiral Howe, his brother, with one hundred and fifty vessels, issued an Order stating:

"The time is now near at hand which must probably determine whether Americans are to be freemen or slaves; whether they are to have any property they can call their own; whether their homes and farms are to be pillaged and destroyed, and they consigned to a state of wretchedness, from which no human efforts will probably deliver them," etc.

BRITISH CAPTURE NEW YORK.

In June, 1776, Washington, to prevent the passage up the Hudson, in addition to guarding the Narrows, directed Colonel James Clinton, commanding Forts Constitution and Montgomery, to protect the fortifications in the Highlands. To Moylan was entrusted the work of sinking obstructions and erecting a *chevaux de frise* to still further debar the approach of the enemy. Affairs were in a critical position. The Americans were in a weak condition for lack of arms, while every man was needed to ward off the General and Lord Howe's impending blow—an attack on New York. But it is the movements of Moylan that concern our purposes.

On June 29th Henry B. Livingston wrote to Colonel James Clinton:

"The [British] Fleet is arrived at Sandy Hook about 46-sail, and I go to my station to-morrow morning; it is a very honorary one, and one that I am much pleased with as the Genl has granted me great power. Col'l. Moylan tells me that a Number of stores are ready to be sent to the Forts."

On July 4, 1776, Dr. James Clitherall of Charleston, S. C., records that on that day he called on General Washington who invited him to dine with him, "but being engaged with Colonel Moylan, Quartermaster-General of the American army, we could not accept of it." He dined with Washington the next day. The Doctor left Philadelphia on July 2d, "the glorious day that threw off the tyranny of George III," carrying letters of introduction to Washington, Mifflin, Putnam, Moylan, Reed and Dr. Morgan.

On July 16th Congress Resolved: That the Secret Committee be directed immediately to deliver to Mr. Mease to be sent forward to Colonel Moylan, Quartermaster-General, half a ton of saltpetre.

Colonel Moylan, at New York, July 28, 1776, wrote to Jeremiah Wadsworth, New Haven, Conn.: "Hoped to have heard the night before of success to the eastward; their folks are idle for want of vessels; asks him to send fifty barrels of pitch and of turpentine, also coal; the stoppage of the North River deprives them of plank and scantlin; asks him to send what he can pick up of these."

OBSTRUCTING THE HUDSON.

At this time Colonel Moylan was engaged in sinking a *chevaux de Frise* opposite West Point on the Hudson to debar the passage of the British up the river; he wrote to Committee of Safety at Philadelphia for one fitted to manage the erection of the work. The Committee replied that there had been "a proper Person spoke to to superintend the Water Chevaux de Frise at New York. It will be known to-day whether he will undertake it, if not, some other person will be immediately sent to you. In the mean Time you may provide the logs and engage the Workmen, as very few, perhaps not more than two or three, can possibly be spared from the Works here. In this Business House Carpenters who may be found among the Troops may be employ'd as well as Ship Carpenters." "Some suitable Person will be sent you to rig the Gallies and sink the Chevaux de Frise." [*Penna. Arch.,* 1st Series, 1776.]

Later the Committee wrote: "The bearer, Mr. Arthur Donaldson, is a person of Good character and has perfect skill & knowledge in constructing those kind of Machines."

SUPPLIES GATHERED BY MOYLAN.

In July, 1776, British vessels passed all the American batteries without injury up the Hudson. Washington believed this was intended "to cut off all intercourse between New York and Albany and to prevent supplies from coming or going, but he wrote Congress, "that the commissary has told me that he has forwarded supplies to Albany sufficient for 10.000 men for four months; that he has a sufficiency here (New York) for 20,000 men for three months and an abundant quantity secured in different parts of the Jerseys for the Flying Camp, besides about 4,000 barrels of flour in neighbouring parts of Connecticut. Upon this head there is but little occasion for any apprehensions, at least for a considerable time." [*Sparks'*, III, 476.]

This shows the efficiency of Colonel Moylan in gathering and transporting supplies, though there came difficulties later.

On August 12, 1776, Washington sent Congress the "Return" of stores sent by Moylan to General Schuyler at Albany and a week later a report of the supplies sent General Gates. [*Am. Ar.*, Vol. I.]

On August 22, 1776, Colonel Clement Biddle from Perth Amboy to Colonel Moylan at New York: "I last evening removed about 60 bbls. corn from the wharf (which has layn there from the morning for Mr. Marsh) to the barrack as we had information of an attack on this place being intended but it proved false. I have several matters on which I must entreat your directions but the time is too full with expectations of great events to take off your attention at present. Last evening the Commissary informed me he had that day issued 5400 rations and 400 R's retained at this port of Perth Amboy but Col. Griffin tells me he cannot by the returns to him conceive there are 4000 men here; they are continually coming in, chiefly militia; the flying camp forms slowly owing to all the militia being ordered out, which occasions our having a disorderly tho a fine body of men."

Lord James Drummond, claiming authority to arrange terms of conciliation, endeavored to gain interviews with Washington who would in no manner recognize him. He was arrested and paroled—went to West Indies and returning sought Lord Howe's permission to land in New York. He, on August 19, 1776, wrote Washington in reply to his of 17th: "I had taken the precaution to prepare a letter to Colonel Moylan on that subject and which I read to Mr. Tilghman on his delivering me that of your Excellency, but which I forbore delivering as not thinking it sufficiently explicit."

He sent his letter to Moylan and desired an interview to "afford me an opportunity of exculpating myself or place me in the situation to suffer that treatment which follows an infraction of parole." [*Wash. MSS.*, VI, 179.]

On August 27, 1776, was fought the battle of Long Island so disastrous, especially to the Maryland Line, to Washington's army. This was followed by Harlem Heights, September 16; White Plains, October 28th; and the capture of Fort Washington by the British on November 16th, all more or less unfavorable to Washington and necessitating the memorable and marvelously strategic retreat through New Jersey.

On September 1, 1776, Colonel Richard Carey, Aide to Washington and by his order wrote General Heath notifying him that Washington had directed Colonel Moylan to furnish the horses wanted and that Moylan was to place an assistant Quartermaster at King's Bridge to supply such articles as are necessary.

On the 3d Colonel Moylan from New York wrote General Heath at King's Bridge that Washington had "ordered the tools and necessaries to some place of security from the enemy and as convenient as the situation will admit of, to Heath's encampment. Major Bacon was sent to consult and take Heath's orders.

The same day Moylan was notified by Secretary Harrison that the Committee at Albany had sent boards to Peekskill and that Washington desired Moylan to get them down to King's Bridge or some place near it and that he would exert himself to have "a pretty considerable quantity provided as many will in all probability be wanted to shelter the troops that may be stationed there and at the posts about it."

On the 6th Colonel Moylan wrote John Hancock, President of Congress, concerning a quantity of Russia duck, which was in the hands of "Thomas Greene, Esq., of Providence, which is ordered by the Secret Committee to lay till further orders from them. We are here in great want of tents. If you could procure an order from said committee to have the duck made up into tents and forwarded to me it would be of the greatest service to the army."

September 9, 1776, Washington's order to Moylan was "that you would without loss of time set about preparing a Sufficient Quantity of Boards, Scantlin and every Material necessary for the Building of Barracks at King's Bridge and the posts thereabouts. The North River, down which most of the Articles must come, is now entirely free from any Obstructions by the Enemy, but how long that may continue is uncertain. The Season advances fast, when it would be impossible for the Troops to lay in Camp, even if they were all supplied with Tents and had a sufficient stock of Blankets and other warm Cloathing; but you well know that in the Article of Tents, at least one-third part of the Army are unprovided, and those that we have are worn and bad; as to bedding and other Cloaths they are in a manner destitute. We have every reason to fear and suppose, that the great naval Force of the

Enemy will oblige us to quit this City whenever they please to make an attack upon it. We must then depend upon Barracks for Shelter, and for that reason you and your deputies to exert yourselves, in the most strenuous manner, in collecting such a stock of wood for the Buildings, and Brick or Stone and Lime for the Chimnies and ovens, as will enable you in a short time to provide comfortable coverings for the men, at the different posts."

CONGRESS INVESTIGATES CONDITION OF ARMY.

Congress, on September 21, 1776, appointed Roger Sherman, Elbridge Gerry and Francis Lewis a Committee to "inspect the state of the army at New York." They left Philadelphia the same day, arrived at Washington's camp on 24th and spent three days in the examination. They reported to Congress on October 2d that the number of men in the army was 25,375 of which 16,905 were fit for duty, 1543 on command and the residue sick or absent; that the militia on their march to the camp, but not included in the report "would amount to upwards of 11,000 men"; that the army was well supplied with provisions, except vegetables which were not then to be procured; there was a "want of salt"; the sick have been greatly neglected and numbers have "dyed for want of necessaries and attendance"; military stores are wanted; the military chest has been too frequently unsupplied with money, but at present the paymaster has a sum fully equal to the General's wishes; that cloathing and blankets are greatly wanted and a supply has been neglected as well from the want of a proper officer to superintend the business as from the scarcity of these articles; military discipline did not prevail.

MOYLAN REQUESTED TO RESIGN.

How and to what extent these delinquencies were chargeable to Colonel Moylan does not specifically appear, but it is evident from a letter of William Ellery, Delegate of Rhode Island, written September 27th to Governor Cooke of that Colony, that Colonel Moylan was held responsible for sufficient to warrant the Committee to ask him to resign. Ellery wrote:

"The Committee who were appointed to inspect the state of the army at Harlem have returned and represented things in a more favorable light than we had used to view them. Methods are taken that the army shall be better disciplined and provided in every respect than it hath been. Although we have some good officers in some of the principal departments, yet in others there is great want of skill and abilities. The Quartermaster-General, Moylan, was persuaded by the Committee to resign, and Brigadier-General Mifflin was persuaded by the Committee to accept that office, with the rank and pay of Brigadier-General. This appointment will give great satisfaction to the army, for General Mifflin is not only well acquainted with the business of the office, but he hath spirit and activity to execute it in a proper manner." [*Journal of Congress*, V, p. 844; *R. I. in Rev.*, 89.]

MOYLAN'S EXPLANATION.

Colonel Moylan's letter to Congress gives an explicit relation of the conditions of the army at the time and of the difficulties he encountered. It was addressed to the President of Congress:

HARLEM HEIGHTS, 27th Septr 1776.

SIR:—

The Field Deputies from Congress Conferd with me this day on the business of the Quarter Master Generals department, they told me that they found a disatisfaction prevail in the Army, by its not being Supplied Sufficiently with the necessaries in that Department, That it was their wish, to reconcile a body of Men So very necessary for the defence, of the glorious Cause we are all engaged in, and proposed that General Mifflin Should resume that department as it appeared to them an effectual method, of giving Satisfaction to the Army, and bringing the department into more regularity, which I must own, there has of late been great need of oweing to Causes, which I shall take the Liberty of pointing out to you, and through you Sir to Congress before I close this.

These Gentlemen urged the necessity of this plan, which they had adopted, So forceably, and at the same time, in So delicate a manner, that I did not hesitate, in telling them, that as a Servant to the publick, I would very willingly resign my office, as it appeared to them, to be for the publick good. I placed my houour in their hands and I shall be very much mistaken in them, if they do not treat it with tenderness. They were pleased to tell me, I might have the Command of a Battallion, which tho I hold to be a most honourable post, for the following reasons I have declined:

1st As the Quarter Master General is at the head of the Staff, I conceive that he takes rank of all Colonels in the Army, it being generally the Custom in most Nations, to give the Rank of Colonel to the assistants QrMrGenl at the end of one or two Campaigns, I therefore think that it would be going back in the army, rather than advancing which is what I can not reconcile to my feelings, especially, as there is a precedent, which differs.

2dly Tho I have employed my Spare time, in Studying the art of War and for fifteen months past have seen a great deal of its practice, My vieus were turned to the Grand and extended parts thereof more than to the Minutio. I do not therefore think myself Capable of teaching a new Regiment the necessary duties. These Sir are the reasons by which I am actuated. At the Same time I can assure the Congress, that I am very willing to Sacrifice my Life, when Calld upon, in the glorious Cause which from the noblest principle, I have voluntarily engaged in. I shall settle my accts with the Commissioners as Soon as possible, and Serve a Volunteer in this Army, untill Congress is pleasd to point out Som other Line of Duty for me.

I will now Sir beg Leave to mention the principal Causes which have given rise to the disatisfaction in the Army with my Department. When I had the honour of being appointed to the office, the Navigation of the North and the East Rivers, were ours, every thing wanting was Conveyed to us by these Channels every thing went on Smooth, easy, Well. The few Waggons, and horses we had, tho allmost wore down in the Service, with a few more added by me, were Sufficient for all the exigencies of the Army, it was a long time after the arrival of the enemy before there was any just Cause for Complaint. A Large part of our Army was detached to Long Island, Waggons, Carts, and Horses were necessary to be sent over, many were sent thither. Perhaps there does not occur in History a Sadder retreat, so well Concerted, So well executed, than was made from that Island. But our Waggons, Carts, and Horses, Could not be brought on, the Navigation of both Rivers was stopped. Of course we were deprived of our usual Supply and then Complaints began, we wanted Waggons to do that duty, which boats were accustomed to do.

I used every endeavour in my power to remedy the evil, it was too Sudden, and not in the power of man to provide time enough for the emergency.

The Cooking utensils of Many Regiments Left on the Island, the fluctuating State of the Militia Coming in destitute of every necessary, drained our stores, and it must take up time to get fresh supplies. To this I may add, demands upon the Quarter Master General, before unheard of, in any army, which not being Complied with, gave Cause of Complaint.

The removeing of the Stores from New York very Soon Commenced, all our own and all the Teams, that Could be pressed or hired, were employed in that important Service, the Commissary Generals, the Director General of the Hospital, the Commissary of Artillery, and what Stores remained in my department, must be sent off and that Suddenly, this Movement naturally alarmed the army in and about the City, they wanted teams to move their baggage, &c., and none Could be Spared. This Caused great Clamour and the QrMr General must be to blame, the Stores of the different departments were Crowded promiscuously on board of every vessel and boat we Could procure, no Store houses to put them in, provided. Of Course Confusion in the extreme, did ensue, I may be asked why Storehouses were not provided, the maneuvre was unexpected no time allowed to build and very few houses or barns in this part of the Island.

We were just emergeing from this Chaos when the field deputies arrived, but the Clamour of the Army had not time to Subside, the Loss of Baggage which were Loaded on Waggons, all falling into the enemys hands, irritated them, and I do suppose the representations of many, were strong against me, tho themselves were chiefly to blame for Leaving their Baggage in their great hurry. The Deputies from Congress were alarmed at the many Complaints and proposed, the remedy, which I chearfully acquiesd in, as it was their opinion, that it would be for the good of the Service. General Mifflins abilities were tried in this Department, they are great, and I Sincerely hope he will reconcile all matters, the provision I have made will assist him greatly. Timber, plank, boards, nails, brick and Lime are engaged in Sufficient quantitys to build Barracks for the Army, I have contracted for ten thousand Camp-kettles which are daily comeing in. Fifty Waggons with four horses to each, are now purchasing in Pensilvania, there are between this and Norwich Comeing to Camp fourteen thousand Canteens and a Large quantity of pails, with many other articles which would take up too much of your time to enumerate. I must beg pardon of you Sir and the Congress for taking up so much of it as I have done, but justice to my own Character, will I dare say, with Gentlemen of Your Liberal Minds plaid my excuse.

I will therefore add no more, than assureing you and them, that I am with the greatest respect

Sir Your Most Obliged
and very Humb. Sert.

STEPHEN MOYLAN.

[*Papers of the Continental Congress,* 78, Vol. XV, pp. 101-108.]

Elbridge Gerry wrote to General Gates on September 27, 1776: "We have obtained Colonel Moylan's resignation and General Mifflin comes again into the office of Quarter Master General."

Washington's General Orders of September 28th, issued at Harlem Heights, announced:

"Stephen Moylan, Esq., having resigned his office of Quartermaster General, Brigadier General Mifflin is appointed thereto till the pleasure of Congress is known."

In "the latter part of September," 1776, James Allen visited the American army at Fort Constitution or Fort Lee, lodged with Washington and "found there Reed, Tilghman, Grayson, Moylan, Cadwallader and others and was very happy with them." [*Pa. Mag.,* 1885, p. 192.]

Cæsar Rodney, writing to Thomas Rodney, Philadelphia, October 2, 1776, said:

". . . General Mifflin came to town the day before yesterday He brought letters from General Washington informing Congress that Mr. Moylan, the Quartermaster General, had resigned his commission, as unable to conduct the business of so many troops. That in consequence thereof, the General had prevailed on General Mifflin to accept, confident that there was not another man in the army who could carry on the business upon the present large plan."

John Jay, writing from Fishkill, New York, to Edward Rutledge, October 11th, said: "Moylan acted wisely and honestly in resigning."

MOYLAN A VOLUNTEER.

Moylan "remained constantly with the army as a Volunteer," wrote Washington to Congress, January 22, 1777.

MIFFLIN SEEKS SUPPLIES.

General Mifflin, on September 28, 1776, applied to the New York Provincial Congress, through Captain Berry, for supplies of lumber, shingles, wagons, horses and men, it being of the utmost importance to the preservation of the army and the general cause of America that these should be procured with the utmost expedition.

THE COMMITTEE ON ARMY.

Congress, on October 1st, reported that three hundred thousand dollars were necessary "to enable the new Quarter Master General to supply the various and necessary supplies."

That would suggest the main lack of Moylan—" dollars necessary to enable" him "to supply the supplies."

Washington, forced to abandon the defense of the Hudson and protect New York, retired to New Jersey and there began that wonderful strategic movement which marked him as a man of military abilities and reserved power for critical occasions. Backward and backward toward the Delaware until, in bleak and chilly December, he had been pushed across the river. Suddenly, on that drear and wintry Christmas night, he recrossed the Delaware with his little army and—Trenton was fought and won.

MOYLAN A VOLUNTEER AIDE.

As volunteer Aide to Washington, Moylan, from Morristown, on December 15, 1776, wrote to General Heath: "My business was to push Lee's and Gates' armies to join General Washington. What the General gave me in command was to proceed to General Lee and Push him forward. The day I came up with his army was unfortunately the day he was taken prisoner. I was to proceed with him to push Greaton's, Bond's, Porter's, Patterson's, Stark's, Poor's and Reed's. I will follow if I can find out their route."

Two days later General McDougal reported to Heath: "I found Colonel Moylan here [Morristown] in quest of the troops of General Gates in order to quicken their march to join General Washington."

Pressed by the British, Washington reached the Delaware and crossed to the Pennsylvania shore. All seemed lost. "I am at the end of my tether. In ten days this army will have vanished," he said. "The game is up," he wrote his brother on December 18th, and as late as the 23d repeated the same disheartening cry.

The British, too, were certain the end had come. "We have the old fox at last." Cornwallis had his baggage aboard ship at New York intending to return to England.

The "Rebellion" had been conquered. The freezing of the Delaware and the expiration of the enlistments of the greater part of Washington's army alone delayed the crushing of the "foul and unnatural" revolt. Washington told Robert Morris: "You might as well attempt to stop the winds from blowing or the sun in its diurnal as to stop the men from going when their time is up"—January 1st.

But Washington awaited not the freezing of the water nor the expiration of term of enlistment. Washington's army had been reduced to about 2400. He sent hastening messages to Philadelphia to push on the militia. Captain John Barry and Thomas FitzSimons, Colonel Moylan's brethren in Faith and his fellow-members of the Friendly Sons of St. Patrick, organized companies and hastened to Washington's assistance.

But Washington, admirable in conception, in times of greatest danger, at this—the crisis—would seem to have had the guidance of that Divine Providence on Whom he constantly relied. He projected a forward movement—an attack upon the whole British line from above Trenton down to opposite Philadelphia. Colonel Joseph Reed was entrusted with operations to cross from Bristol, Pennsylvania. He went to Philadelphia on the evening of December 24th—Christmas Eve—to persuade General Putnam, who commanded at Philadelphia, to send a force across the Delaware to Cooper's Ferry. There he met Colonel Moylan who had been sent by Washington to hurry on supplies. Robert Morris, writing on December 23, 1776, to President Hancock of Congress, stated:

"I received last night a letter from General Washington, per Col. Moylan, requesting me to hurry Mr. Mease, to have soldiers' clothes made up with all possible diligence. He says muskets are not wanted there, but that comfortable clothing is exceedingly wanted. Colonel Moylan advises by all means to send up the stockings and great coats, now arrived, which I think I do." [*Bull. Pa. His. So.*, Vol. I, 1845-47, p. 58.]

"I am informed by Mr. Moylan that Col. Guyon (I think that is the name) was taken prisoner with General Lee."

The "Andrew Doria," Captain Joseph Robinson, had just arrived from St. Eustatia, capturing a British supply sloop and bringing her to Philadelphia. She had stockings, jackets, coats, blankets, linen, muskets, powder and lead. [*Ibid.*]

Six years after this time—in 1782—Reed, when President of

Pennsylvania, was charged by Dr. Benjamin Rush and General Cadwalader with, at this critical period in affairs, having contemplated going over to the British and with actually, at Burlington, having taken a protection from Colonel Donop the Hessian.

In his defense he thus relates his presence in Philadelphia on Christmas Eve:

"I lay down for a few hours and when morning came a number of gentlemen, among whom I particularly recollect Colonel Moylan, Mr. James Mease and Mr. R. Peters, came and anxiously enquired into our situation and prospects. They can tell whether despondency or animation, hope or apprehension, most prevailed, and whether the language I held was not the very reverse of despair; the former [Moylan] may remember, that when urged to stay and partake of a social entertainment provided for the day, I declared my resolution that no consideration should prevent my return to the army immediately; and that in a private conversation I pressed him to do the same, lest he should lose a glorious opportunity to serve his country and distinguish himself. I was not at liberty to be perfectly explicit, but the hint was sufficient to a brave officer." [*President Reed,* by Wm. B. Reed, 2d Ed., 1867, p. 106.]

This "brave officer"—Moylan—was one of Reed's life-long friends.

Bancroft's *History of the United States* in describing the severity of the weather this Christmas Day, states: "Moylan who set off on horseback to overtake Washington and share the honors of the day became persuaded that no attempt could be made in such a storm and stopped on the road for shelter." [IX, p. 229.]

Bancroft reiterated the charge against Colonel Joseph Reed whose grandson, William B. Reed, made vigorous defense, as his grandfather had done at the close of the Revolution. Several pamphlets were issued on the subject which excited great interest forty or more years ago. As an illustration that "Truth will prevail," it must be noted that it was not until 1876 when General Stryker, of New Jersey, whose Revolutionary historical narrations are of the highest import and value, settled the controversy by presenting evidence that it was Colonel Charles Read of the New Jersey Militia who had the interview with Donop, took "a protection" from him and abandoned the American cause.

Yet President Reed's enemies in his lifetime and Bancroft in our days recited, reiterated, argued and presented plausible proofs of the despondency of Reed and of his intention to accept British allegiance, but that the victory at Trenton had caused him to change his mind. Yet it was all a case of mistaken identity.

It was just one hundred years after the alleged intended desertion that evidence was found vindicating Reed's memory even to the satisfaction of Bancroft, who, when presented with the proof, requested that he be allowed to be the first to make it public. General Stryker consented thereto.

Then came the Battle of Trenton when, in a terrific snowstorm and keen, cutting cold, Washington crossed the Delaware and, attacking by daylight in the morning the Hessians, achieved a victory which gave new hope, new life to the Patriots and changed the whole aspect of affairs, followed as it was by the victory at Princeton a week later.

The day after the Battle at Trenton—December 27th—Colonel Moylan, at Newtown, Pa., wrote Robert Morris, at Philadelphia, sending a Return of the Prisoners taken at Trenton. They were: 1 Colonel, 2 Lieutenant-Colonels, 3 Majors, 4 captains, 8 Lieutenants, 12 Ensigns, 92 Sergeants, 9 drummers, 25 officers' servants, 740 rank and file. Total, 918. This said Moylan was "a rough but a just sketch as I can collect. I was unfortunately too late to share in the honours of the day, being catched in the storm and little imagining that any attempt would be made at such an inclement time. Our loss very inconsiderable. Six pieces of artillery, four standards, one thousand stand of arms were taken. If the whole plan could have been put into execution there is little doubt but the whole of the Hessians along the Delaware would have been done for.

The spirited conduct of the whole who went over did not exceed twenty-two hundred was great. You must remember what a morning yesterday was for men clad as ours are, to march nine miles to attack an enemy provided with every necessary and elated with a succession of advantages over our handful of men whom they were accustomed to see retreating before them. [*Am. Ar.*, Vol. III, 1446.]

THE BATTLE OF PRINCETON.

But if Moylan "was too late" for Trenton he was on time for Princeton. How he must have conducted himself in that brilliant manœuvre of Washington's is attested by his letter to Robert Morris.

AMERICA MUST BE FREE.

How jubilant! How our hearts beat joyously as we read:

HEADQUARTERS AT MORRISTOWN, 7th Jan'y, 1777.

DEAR SIR:—I thank you, my good friend, for your favor of the first. What a change in our affairs, since the date of that letter. Are you not all too happy? By Heavens, it was the best piece of generalship I ever heard or read of. An enemy, within musket-shot of us, determined and only waiting for daylight to make a vigorous attack. We stole a march, got to Princeton, defeated, and almost totally ruined three of the best Regiments in the British service; made all their schemes upon Philadelphia, for this season, abortive; put them in such a consternation, that if we only had five hundred fresh men, there is very little doubt but we should have destroyed all their stores and baggage, at Brunswick, of course oblige them to leave the Jerseys (this they must do), and probably have retaken poor Naso. What would our worthy General have given for 500 of the fellows who were eating beef and pudding at Philadelphia that day? But let us not repine—it was glorious. The consequences must be great. America will—by God—it must be free.

I never mentioned my desire to the General of engaging in the Cavalry. Your letter, I believe, gave him the first intimation. I put it into his hands to show your gift of divination. Pray how could you suppose, that our next blow must be at Princeton, but I recollect you did not then know we were attacked at Trenton. How your heart went pitipat, when that news reached you, and what an agreeable feeling you all must have had when you heard of their facing right about. But that feeling is very short of those which we all enjoyed when pursuing the flying enemy. It is unutterable—inexpressible. I know I never felt so much like one of Homer's Deities before. We trod on air—it was a glorious day. Pray send us back those runaways that left us these some days past. We are really weak—strengthen our hands, and we will not leave an enemy out of gunshot from their ships. I will not tire you farther than telling you what I have often done, that I am sincerely, Sir, Yours,

To Robert Morris, Esq. STEPHEN MOYLAN.

[*Pennsylvania*, August 30, 1855; Reed's *Pres't Reed*, 2d Ed.]

"Naso" meant General Charles Lee. He had been captured by his own connivance it is now suspected, as later actions seem to prove.

Washington to Congress, 22d January, 1777:

"Colonel Baylor, Colonel Moylan (who, as a volunteer, has remained constantly with the army since his discontinuance in the Quartermaster's Department) and Colonel Sheldon, command the three new regiments of light dragoons." [*Sparks,* IV, 293.]

APPOINTED TO COMMAND CAVALRY REGIMENT.

As we see by Molan's letter, Washington had prior to January 7, 1777, and probably in December, selected him to organize and command a regiment of cavalry. His commission, however, dated from January 8, 1777. On January 21st, Washington so informed the Congress, and Moylan was commissioned to recruit a Light Horse Regiment. He went to Philadelphia to seek recruits and select officers to do so in Maryland.

On 2d February, 1777, writing to Washington, then at Morristown, he said:

"The 2000 Dollars which I received your warrants for, is dispersed amongst the officers some of whom have got the horses for their Troops, and make further demands upon me; if you give me an order on the Committee of Congress for what I may have occasion, for the completeing the Regiment, it will save the trouble of warrants, & spare the military chest."

MONEY TO MOYLAN.

The Resolves of Congress relative to Moylan's Regiment may be summarized as follows:

On February 6, 1777, Washington, at Morristown, wrote to the Committee of Congress requesting money for the recruiting of Moylan's Regiment. A copy of this request in the handwriting of John Fitzgerald, Washington's Aide-de-Camp and a fellow-Catholic with Moylan, is among Washington's Papers in the Library of Congress.

February 26, 1777. Ordered $3000 to be paid "for the services of recruiting his Regiment."

On April 8, 1777, Congress ordered two warrants of $12,000 each to be issued in favor of Colonel Moylan. On May 16th an order for $10,000 more was ordered, and on May 30th $25,000 additional was ordered. He was directed to lay before the Board of War an account of the expense of raising and equipping a troop of light horse.

Colonel Moylan wrote General Washington, at Morristown, from Philadelphia, 14th April, 1777:

"I had fixt on this day, for my setting out for Maryland, but the hostile appearances in this Bay are such, that I have given up the thought of going there, and have wrote to Major Washington to repair to Baltimore, and take the part of the Regiment raiseing under his command, I have recommended him to push them forward as soon as possible, which I dare say he will be as anxious in doing as I shall be. Considering the circumstances of not having a place to train either men or horses, during the bad weather, I have the pleasure to inform your Excellency, that the part of the Regiment here are pretty forward in their exercise. I have them out every day, and if the enemy will give us time to have them properly equipt, I flatter myself with the thought, that the Regiment will not disgrace our arms. Mr. Mease promised me the Regimentals of one of those that were taken, from the enemy. He now tells me, there have been so many applying that if I have not your sanction, he doubts much whether I shall be able to get it."

THE RED REGIMENTALS.

These "Regimentals that were taken from the enemy" caused consternation and alarm when worn by Moylan's troopers. So much so that General Washington was obliged to order the color to be changed, as he set forth very clearly in this letter:

"HEAD QUARTERS, MORRIS TOWN, 12 May, 1777.

"A party of your Regiment arrived here Yesterday with an escort of Money. Their appearance has convinced me fully of the danger which I always apprehended from the similarity of their Uniform to that of the British Horse, and the Officer who commands the party, tells me, that the people were exceedingly alarmed upon the Road, and had they been travelling thro a part of the country, where it might have been supposed the enemy's Horse would be foraging or scouting, they would in ail probability have been fired upon. The inconvenience will increase when your Regiment joins the Army. Your patroles will be in constant danger from our own scouting Parties and when ever there is occasion to dispatch a party into the country, they will alarm the Inhabitants.

"I therefore desire that you will immediately fall upon means for having the colour of the Coats changed, which may be done by dipping into that kind of dye that is most proper to put upon *Red.* I care not what it is, so that the present colour be changed."

In the campaign which followed, Moylan's Light Dragoons wore green coats trimmed with red, green cloaks with red capes, red waistcoats, buckskin breeches and leather caps trimmed with bear skin. [Mellick: *Story of a Farm,* p. 463.]

"DRAGOONS."

The title "Dragoons" which, now for the first time, we find applied to Colonel Moylan's Light Horse, and that, too, by General Washington, has been the subject of historical consideration as to its origin. Dr. H. C. Parry, U. S. A., in *The United Service* for August, 1881, says, in speaking of the troopers in France during the time of Louis XIV, in 1685:

"We are justified in believing then the term Dragoon was derived from the cruelty and devastation caused by these troopers being compared to the pernicious qualities the dragon was supposed to possess and the evils he inflicted on mankind."

The Light Horse Regiments at this time were Sheldon's (recruited in Connecticut), Baylor's, Moylan's and one in Virginia. All were to join Washington's Army at Morristown for the campaign of 1777.

When the four were joined at Morristown a proposition to commission a General of Horse was considered by Congress.

On May 24, 1777, Washington wrote Moylan, then in Philadelphia, that if Congress left the appointment with him he would name [Thomas] Reed, who had lately been named as Brigadier-General by Congress. He requested Moylan to so inform Reed and to assure him of the pleasure he had had in learning of his appointment as Brigadier. [Ford, V, 389.]

But Reed appears to have been made Paymaster-General.

COURTS-MARTIAL.

On May 21, 1777, General Schuyler laid before Congress the proceedings of a Court-Martial, held at Philadelphia on Monday and Tuesday, May 19th and 20th, on the trial of John Brown, alias John Lee, signed by Stephen Moylan, President, wherein the Court determined: "That the prisoner is guilty of conducting five men to Brunswick; of holding a traitorous correspondence with the enemy, in offering himself as a pilot to General Howe, to conduct the British army from Brunswick to Philadelphia; and also in promising to discover to the enemy to what place the continental stores, from Philadelphia, were removed; and the Court found the prisoner guilty of a breach of the nineteenth section and were of the opinion that he should suffer death; but, that from some circumstances which appeared on his trial, the Court think proper to recommend him to the General as an object of mercy.

"Ordered, that the said proceedings be referred to the Board of War; and that they enquire into the circumstances that induced the Court to recommend the criminal as an object of mercy and report to Congress."

In Congress, May 23, 1777, the Board of War brought in a report which read:

"That the Board have conversed with Colonel Moylan, the President of the Court-Martial held on John Brown, now under sentence of death as a Spy and Traitor and it appears from Colonel Moylan's Report, that previous to the sentence of the Court, after examination of witnesses, two of the members were sent to the criminal to endeavour to find out his accomplices. That he mentioned several persons in Northampton County, in Pennsylvania, who appeared to be his Relations and connexions and whose names were ordered to be given to General Schuyler: but as to his case no particular circumstances of mitigation appear except such as are founded in the ignorance of the culprit. The reason for a motion in the Court-Martial for a recommendation to mercy was founded only on the criminal's apparent ignorance and illiteracy."

MOVEMENTS OF THE ARMY.

At this time, May-June, 1777, Washington was at Middlebrook, New Jersey; General John Sullivan was at Rocky Hill and Moylan's Dragoons, recruited in Philadelphia and vicinity, covered the region round about Middlebrook,—at Woodbridge and Spanktown.

In the morning of June 14th the enemy, said to have been 2000, advanced to Van Ests Mill and skirmished with Colonel Daniel Morgan's riflemen on their way towards Princeton, seemingly moving again towards Philadelphia, though this was a feint as the British later sailed southward from New York to the Head of Elk, Maryland, and came northward.

Washington, on June 13th, sent Charles Thomson, Secretary

of Congress, the draft of a plan for the establishment of cavalry The next day he sent the Board of War "the establishment and pay of light dragoons."

General Reed wrote Washington from Philadelphia, 18 June, 1777, saying:

"Colonel Moylan writes that he thinks my knowledge of the country and of the people would be of use in the quarter where he is, and presses me to come up, which I shall do immediately." [*Cor. Rev.*, I. 389.]

Washington remained at Middlebrook until July 12th, when he established his headquarters at Pompton Plains, New Jersey, where Colonel Moylan wrote him as follows:

MOYLAN SEEKS PROTECTION FOR A QUAKER PHYSICIAN.

"BLANDS, 12th July, 1777.

"I lodge in the house of a man who I verily believe has as good a heart as any man can have, I have had many opportunitys of enquireing into his character. He lays out at least one hundred pounds sterling p. annum amongst the poor of his neighborhood; he is a Phisyscian who receives no fees, he is the friend of the distress'd—this gentleman is by proffession a Quaker of course a peaceable man, he has relieved many of our Soldiers in their need, he has entertained our officers who all have a good opinion of him, it is true he has entertained General Howe & his officers, and I believe from the same principle. he has suffered by both armys, but not by the gentlemen of either army, he knows how far your name will influence even the common soldier engaged in our cause, and wishes for a protection under your hand, I have promised him that I would apply for it, and did not doubt that if your Excy. saw no impropriety in it, you would give my friend Elias Bland a protection for himself, familly & effects as far as you with consistency could do it."

RATIONS FOR MOYLAN'S MEN.

The annexed requisitions for rations show the strength of the companies commanded by Captains Dorsey, Hopkins and Plunket of Moylan's Regiment.

Woodbridge [N. J.], 16th July, 1777. Capt. Thos. Dorsey's troop of Col. St. Moylan's Reg't drew 52 Rations for 1 Capt., 1 Lieut., 1 Cornet, 1 Quartermaster, 2 Sergeants, 1 Farrier, 2

Waggoners and 6 women, 30 Rank & file. Total 53 [Less] Retain 10 equals 43. Signed John Craig, L. [Light] Dragoons.

July 16, 1777. Capt. David Hopkins, troop drew 55 rations.

Capt. David Plunkett's Troop at Spanktown, July 16, 1777, drew 54 rations.

[Original of above requisitions were in lot 931 of "Letters and Documents relating to Colonial and Revolutionary times." Sold at Davis & Harvey's, April 3, 1906.]

MOYLAN REPORTS TO WASHINGTON.

Colonel Moylan reported to Washington, who was "near Smith's Cove, New York," from Elisthtown, 21st July, 1777, that he had received orders to march to New Windsor, taking my route by Clone, which I understand is near ten miles round, however I will obey it except I meet contrary orders, I should have been further on the march had not an unlucky thought taken hold, last night, of nineteen of Craig's Troop, who set off from hence towards Philadelphia at twelve o'clock. Colonel White and the Major with two troops brought them back, after a very severe chase of upwards of forty miles. I believe they have not been well used in regard to pay, which they give as the reason of their proceeding towards Philadelphia in order to get a settlement. I shall not however trust them on horseback again, and when an opportunity of a Court-Martial offers, they shall be brought to trial. The horses they rode, and the horses rode by their pursuers are so stiff that I cannot pretend to move them this day. I hope it will not be attended with any ill consequence as I am informed the enemy's fleet are falling down to the Hook.

These deserters were tried, found guilty, sentenced to death but pardoned.

He sent these reports to the General the next day from South Amboy, 23d July, 1777:

I came over this morning early, to this place to observe the motions of the fleet. They set sail about five o'clock in the morning and stood out to sea with the wind at N. N. West; they made so many tacks backwards & forwards that until noon I could not determine whether they realy meant to go out of the Hook or not; they hauled their wind, after they got out as if going to the Northwd. but that may be to weather the Cape. Eleven sail of ships came thro the narrow and went to sea with the fleet. As soon as I get to my Qrs. shall send this to Col. Dayton to forward to your Excellency, I have been delayed upwards of two hours for a boat, which is at the other side.

The following day this report was sent from Amboy, 24 July:

The Letter from General Forman & the other from my correspondent are just come to hand, I send them to your Excellency, who can form a juster opinion of their contents than I can pretend to do, the Letter for General Forman was immediately forwarded to him.

At this time Moylan and Bland's commands were near Bound Brook, New Jersey. On July 25, 1777, Moylan wrote Bland:

"The inclosed letter came just now to hand. I suppose its contents are orders to march for Philadelphia. My route will be through Princeton, crossing the Delaware at Trenton; yours will, I suppose, be to Corryell's Ferry, which will be the shortest, if you move by Trenton. Let me know, and if I reach before you, the boats shall be in readiness to convey you across; if you get before me you will please do the same for me." [*Bland Papers,* I, 61.]

These reports, with others conformatory, led Washington to hasten forward to the southward all his forces. To Colonel Moylan was sent these orders:

ON TO PHILADELPHIA.

Camp at Ramapough, July 25, 1777.

The Enemy's fleet....................having left the Hook and gone to Sea, I am to request that you will immediately repair with your Regiment to the City of Philadelphia and put yourself under the direction of the commanding officer there. You will not lose a moments time, and will order your baggage to follow under a proper Guard.

In July, 1777, Washington, at Morristown, New Jersey, drafted a detailed route to be followed by the several divisions of the Continental Army on the march southward from Morristown to protect Philadelphia from attack of Howe. The copy in handwriting of Moylan is preserved in the *Washington Papers* in Library of Congress:

At this time Lieutenant-Colonel John Fitzgerald, a Catholic of Alexandria, Virginia, an Aide-de-Camp to Washington, sent by Washington's order, directions to Major-General John Sullivan (whose parents were Catholics and yet he declared the Catholic "a cursed Religion") that the army would retrace its march to the Reading Road beyond the Trappe, a village twenty-five miles from Philadelphia and ordered him to move on a line between it and the Schuylkill where Generals Maxwell and Potter would join him and that he "notify Colonel Moylan to move in concert."

ORDERS TO MOYLAN.

On July 26th Washington was at Mr. Lott's at Ramapaugh, N. J., and in the handwriting of John Fitzgerald, sent Colonel Moylan this order:

Shou'd this Letter come to hand before you pass Trenton, it is His Excellency's Orders that you halt there, either until you receive further Instructions from him, or till you have authentic advice that the Enemy have come into the Delaware Bay, in which case you will govern yourself as before directed. If you have passed Trenton, your best mode will be to proceed to Bristol and there act as ordered at Trenton.

WASHINGTON'S MOVEMENTS.

Washington kept on southward with his army. On July 31st he was at Coryell's Ferry, near Trenton. On August 1st at Chester, Pa., where he received the news of the movement of the British fleet. He was in Philadelphia the next day where he remained directing affairs and consulting with Congress until the 5th. The next day he was at the Falls of Schuylkill where he remained until the 10th, moving to Neshaminy, Bucks County, where the army encamped until the 23d at [now] Hartville. When receiving information that the British fleet had arrived in the Chesapeake the army was started on the march southward to meet the enemy.

While at the Neshaminy camp Courts-Martial, held on August 7, 12, 16, of which Colonel Sheldon was President, tried Moylan's men and sent their verdict to Washington.

Washington on August 19, 1777, approved of the sentences in the case of Edward Wilcox, Q. M. to Capt. Dorsey's Troop charged with taking a horse belonging to Colonel Moylan's Regiment, and a trooper with his accoutrements, found guilty and sentenced to be led around the Regiment he belongs to on horseback with his face towards the tail and his coat turned wrongside outwards and that he be then discharged from the army.

Washington approved the sentence and ordered it to be put into execution immediately. [Saffel's *Records,* 331.]

TRIAL OF DESERTERS.

On the case of the deserters Washington in General Orders announced:

George Kilpatrick and Charles Martin, Sergeants; Lawrence Burne and Enoch Wells, Corporals; Daniel McCarty, Patrick Leland, Philip Franklin, Jacob Baker, Thomas Orles, Adam Rex, Frederick Gaines, Daniel Kainking, Christian Longspit, Henry Winer and Nicholas Walner, privates in Colonel Moylan's Regiment of Light Dragoons, charged with mutiny and desertion, and adjudged worthy of death—the Court esteeming the prisoners, except Sergeant Kilpatrick, objects of compassion, and as such, recommended them to the Commander-in-Chief, the General is pleased to grant them his pardon; and the like reasons which led the court to recommend to mercy, joined with others, induces the General to grant this pardon to Sergeant Kilpatrick also. At the same time the prisoners are to consider their crimes are of a very atrocious nature, and have by the Articles of War subjected themselves to the penalty of death. The remission of their punishment is a signal act of mercy in the Commander-in-Chief, and demands a very great and full return of fidelity, submission and obedience in any future military service which he shall assign them. The prisoners are to quit the Horse and enter into the foot-service, in the corps to which they shall be assigned.

Thomas Farshiers and George House, of Colonel Moylan's Regiment, tried by the same Court, being charged with mutiny and desertion, are found guilty, but some favourable circumstances appearing in their behalf, were sentenced to receive twenty-five lashes on their bare backs, and be dismissed from the horse-service; the Commander-in-Chief approves the sentence, but for reasons above referred to, and with the like expectations of amendment, remits the penalty of whipping. They will be disposed of in the foot-service.

Thomas Runnals, of Colonel Moylan's Regiment, tried by the same Court, being charged, is found guilty and sentenced to death; the Commander-in-Chief approves of the sentence but the execution of the prisoner is respited till further orders. [*Ibid,* 332.]

WASHINGTON'S ORDER TO MOYLAN.

On August 7, 1777, Washington, at Roxboro, now part of Philadelphia City, by Colonel Timothy Pickering, sent this order to Colonel Moylan:

"To-morrow morning you would post careful officers with proper commands of men, on all the roads leading from any part of the camp to Philadelphia in such manner as to intercept every soldier who shall attempt to straggle from the camp towards Philadelphia and all such stragglers are to be taken up, secured and brought on after the army. The whole army is to parade to-morrow morning at 5 o'clock and march as soon after as possible to a new encampment—about 9 miles back; but of this movement you will make no mention but to the officers whom you shall detach for the purpose aforesaid; not to them till you deliver their orders when you send them off. The whole body of horse is to bring up the rear of the army for the business of picking up all stragglers." [*Am. His. Rec.*, II, p. 133.]

Samuel Williamson, Chaplain of Moylan's Dragoons, in August, 1777, intending to go to Europe "to prosecute his studies," the Supreme Executive Council of Pennsylvania on the 23d passed a recommendation of him "to the kindness and notice of all Civil and Military officers of States at peace with the country to which he may go." [*Col. Rec.*, XII.]

THE ARMY IN PHILADELPHIA.

On Sunday, August 24, 1777, Washington's army started at four o'clock in the morning from near Germantown and marched through Philadelphia, "going in and marching down Front Street to Chestnut and up Chestnut to the Common," now the site of the City Hall. It was on its way southward to meet Howe's forces coming northward from the eastern shore of Maryland. Washington's design in marching through Philadelphia was "to have some influence on the minds of the disaffected and those who are dupes to their artifices and opinions."

The cavalry (called the Horse) moved in two wings. Bland's and Baylor's Regiments on the right and Sheldon's and Moylan's on the left, 150 yards in the rear of General Maxwell's Brigade. Washington ordered "Not a woman belonging to the army is to be seen with the troops on their march through the City." [Saffel's *Records*, 336.]

Now-a-days it seems somewhat singular to note the many women attached to the British and American armies during the Revolution—many wives accompanying their soldier-husbands. The well-known instance of Moll Pitcher illustrates the frequency with which women "belonged to the army," as Washington stated it.

The next day—25th—the army of ten thousand crossed the Schuylkill River at the Market Street Ferry and that evening encamped at Darby. The next day, Wilmington, Delaware, was reached. Here Washington, taking Moylan, Bland and Baylor's cavalry, reconnoitered the country and learned that the British had landed at Head of Elk River, Maryland, the day before. The first encounter between the opposing armies was on September 3d, when the British advanced against the American Light Horse Cavalry of Moylan and Sheldon but were obliged to retire.

Washington, on September 6, 1777, in a General Order, stated:

"From every information respecting the enemy's designs, and from their movements, it is manifest their aim is, if possible, to possess themselves of Philadelphia. This is their capital object. It is that they last year strove to effect, but were happily disappointed. They made a second attempt at the opening of this campaign, but after vast preparations and expense for the purpose, they abandoned their design and totally evacuated the Jerseys. They are now making their last effort. To come up the Delaware, it seems, was their first intention; but from the measures taken to annoy them in the river, they judged the enterprise that way too hazardous. At length they landed on the eastern shore of Maryland, some little way in the country, but the General thinks they will again be disappointed in their views. Should they push their designs on Philadelphia by this route, their all is at stake. They will put the contest on the event of a single battle. If they are overthrown they are utterly undone—the war is at an end.

"Now is the time for our most strenuous exertions. One bold strike will free the land from rapine, devastations and burnings; and female innocence from brutal lust and violence." [*Saffel,* 336.]

BATTLE OF BRANDYWINE.

On September 11, 1777, the Battle of Brandywine was fought and, by the Americans, lost. They were unable to make that "one bold strike" which Washington believed would have prevented the capture of Philadelphia and have freed "the land from rapine, devastations and burnings."

"The events of the day," declared Washington in a General Order issued at Germantown, on the 13th, "from some unfortunate circumstances, were not so favorable as could be wished." Yet he had "full confidence that in another appeal to Heaven, with the blessing of Providence, which it becomes every officer and soldier to supplicate, we shall prove successful." So he ordered "thirty hogsheads of rum to be distributed" among the soldiers for their "gallant behaviour and bravery"—"one gill a day while it lasts."

"Though we gave the enemy the ground, the purchase has been at much blood—this being by far the greatest loss they ever met with since the commencement of the war," declared Washington to his soldiers in praising their "firmness and alacrity upon every occasion in the course of the battle." [*Ibid,* 341.]

The British at Brandywine were attacked by "the Light Troops" of General Maxwell's Division "who, after a severe conflict, retired."

The principal disaster of the day happened where General Sullivan commanded. Congress recalled him to answer a Court of Inquiry relative to his action at Staten Island. Washington protested against his recall at the juncture of affairs.

MOYLAN TO REMOVE MILITARY STORES.

The duty assigned Moylan's Dragoons after Brandywine are indicated by the orders of Washington issued the same day—September 13th—from Headquarters at the Falls of Schuylkill:

"The removal of our Military Stores from French Creek is an object of great importance. For the purpose, a number of Waggons must be collected as soon as possible; I therefore desire, that you will immediately detach Twenty Light Dragoons under an active Officer without a moments loss of time to assist Mr. Rittenhouse at French Creek, in procuring such a quantity as he may judge necessary, and to render such other services as he (the Officer) may think material. You will instruct the Officer to procure the Waggons in the least exceptionable manner. But they must be had without one instants delay. The Stores must be removed immediately or they may be destroyed."

French Creek is in Chester County. The stream rises in Union Township, Berks County, and enters the Schuylkill River at Phœnixville. The Board of War, from March 25 to April 16, 1777, sent military stores there, a report of which was sent Washington on April 29th. After the Battle of Brandywine and the subsequent movements of both armies prior to the battle of Germantown these stores were in danger from the British.

From Germantown, 13th September, 1777, Moylan was ordered by Washington to "keep the main Body of the Horse at the Barracks upon Schuylkill; you will keep parties advanced towards the Enemy; move especially upon those Roads leading from Brandywine to the Sweed's Ford and the other Fords upon Schuylkill. Considering the distance that we now are from the Enemy, the security of the Army depends upon the Vigilance of your patrols. The reconnoitering Officers often take up a report of the situation of the Enemy from others and not from their own observations. Impress the Gentlemen before they go out, with the importance of the Business upon which they are sent, and let them know that if any accident should happen to the Army from false intelligence they will be answerable. Your own Judgment will point out any other directions necessary to be given upon the occasion."

PHILADELPHIA TAKEN.

The defeat at Brandywine made the entry to "the Rebel Capital" possible whenever General Howe chose to do so. He took possession on September 26, 1777, being received with great joy by the Tories who remained in the City after the Patriots, not in the army, had hastily removed from the City to the country northward and westward.

Though repulsed, Washington and his men were determined to continue active and unembarrassed. So to "share part in every hardship to which his army is exposed" he and his aides "divested themselves of all baggage save blankets."

On September 14th a movement on Swedes' Ford was ordered, in which Moylan's Light Dragoons took part.

At this time Washington's army was in a deplorable condition for want of shoes, stockings and other necessaries. He wrote Congress on September 23d urging prompt relief and saying, "at least one thousand men are barefooted" and had been marching over the hard roads in that condition.

Moylan's daughter, Mrs. Lansdale, of Baltimore, related that after the Battle of Brandywine Colonel Moylan wrote to Robert Morris urging shoes and blankets for his men. This letter, however, is not among Morris's Papers in the Library of Congress.

THE BATTLE OF GERMANTOWN.

Though arranging to give battle, Washington, on October 3d, sent, under flag of truce, to General Howe his dog which had wandered into Washington's camp. It wore a collar with Howe's name on. The next morning, October 4, 1777, Washington made an attack upon the British at Germantown. Though his men showed "spirit and bravery in driving the enemy from field to field, an unfortunate fog, joined with the smoke prevented the different brigades from seeing and supporting one another . . . and some other causes which as yet," Washington in General Order to his Army said, "cannot be accounted for, they finally retreated; they nevertheless see that the enemy is not proof against a vigorous attack and may be put to flight when boldly pushed. This they will remember and assure themselves that on the next occasion a proper exertion of the powers God has given them, and inspired by the cause of freedom in which they are engaged, they will be victorious."

They really had the British whipped but did not know it. Then two days later, Washington ordered: "Buckshot shall be put into all cartridges that shall be made hereafter."

The following day he announced that "the troops fled from victory," but "nevertheless he has the satisfaction that the enemy suffered very severely."

But the surrender of Burgoyne at Saratoga on the 7th compensated for the non-success at Germantown.

COURTS-MARTIAL.

On October 16, 1777, a Court-Martial of Horse Officers of which Colonel Moylan was President was held at his quarters for the trial of all prisoners of the Horse.

A week later—on October 24th—Colonel Moylan was himself on trial before a Court-Martial held at Upper Dublin, of which Colonel Bland was President. Moylan was charged with disobedience of orders of General Pulaski; with "a cowardly and ungentleman like action" in striking Mr. Fulinski, a gentleman and an officer in the Polish service, when disarmed and putting him under guard and giving irritating language to General Pulaski.

The Court "were of the opinion that Colonel Moylan was not guilty and therefore acquit of the charge against him." Washington approved on the verdict on the 31st. [Weedon's *Orderly Book.*]

The controversy, however, did not end with Colonel Moylan's acquittal. In December Lieutenant Fulinski [Zelinski] "unhorsed" Colonel Moylan. General Pulaski, on the 4th, reported the assault to Washington and sent the letter by the dragoon who had witnessed the affair. [Ms. Sabin, 2:22:98; D. & H.]

Captain Craig of Moylan's Regiment with his party of horse took several British Dragoons and several soldiers with their horses, arms and accoutrements. Washington, on November 9, 1777, issued order desiring "Captain Craig, Captain Lee and the other officers who have distinguished themselves to accept his cordial thanks for the enterprise, spirit and bravery they have exhibited in harrassing and making captives of the enemy." [Weedon, p. 123.]

The orders and reports given show, to some extent, the operations of Colonel Moylan's Continental Dragoons.

CLOTHING FOR MOYLAN'S MEN.

The value of Moylan's cavalry in this campaign is best proven by Washington's declaration to General Heath, made while at Valley Forge, that there had been "so many advantages from cavalry in the course of this campaign that I am determined to augment them as much as possible and against the next."

Desiring that Moylan's men might "take the field in a respectable manner," he sent "Captain Hopkins of Moylan's Light Dragoons to Boston to procure clothing and accoutrements for the Regiment against the ensuing campaign," and informed General Heath that "the Captain will send the articles he may get either made up or not as he finds most advantageous to the Regiment." [*Mass. His. Soc. Col.,* 5th Series, IV, p. 81.]

General Heath, at Boston, on April 7, 1778, issued Pay Warrant "for accoutrements for Colonel Moylan's Regiment of Light Dragoons, which General Heath, by the desire of General Washington, is to furnish." [*MS.*]

We cannot follow the movements of Washington's Army and the consequent manœuvres of Moylan's Dragoons, until they settled in camp at Valley Forge, where they suffered the distress that has become known to all.

DOES NOT AGREE WITH WASHINGTON.

While there proposals for "procuring horses and saddles" were referred to Colonel Moylan by Washington, to whom this reply was sent:

CAMP, 24th February, 1778.

Agreeable to your commands, I have with attention perused the proposals for procuring Horses and Sadles for the Cavalry—as you was pleased to desire that I would give you my sentiments thereon, I must beg leave to tell you that I totally disaprove of the measure. It is arbitrary and cannot be attended with the success expected from it. The mode proposed for putting it in execution must counteract its intentions—for however facile it may appear in theory, for an officer to go into a district to find out the disaffected who have got horses, I believe when it is put in practice, it will be found impossible to keep his business a secret. His attempting to purchase their best horses and saddles at a reasonable price will give the alarm, every man will put his horse out of the way or perhaps move him off to the enemy so that when the time appointed for seizing them comes, few or none will be found, the ill consequences of a very arbitrary act will remain and no good effects accrue from it.

If notwithstanding you should think proper to put the plan in execution, I will order the officers of my Regiment upon that duty, tho I am very certain that they would much rather be ordered to charge a superior number of the enemy than go upon this disagreeable service. Indeed it is that sort of duty that in my opinion the Cavalry should be exempt from. They should conciliate the minds of the inhabitants, not exasperate them, as that corps is more immediately in their power than any other; they must be out in small parties, and depend upon them in a great measure for their security. It may be said that those persons from whom horses are intended to be taken are such, from whom no friendship may be expected, but I apprehend that there is a great difference between those Tories made so through fear, and those who are actuated by resentment, the former will not act against us, the latter will.

Why not, Sir, call upon each State for a quota of horses, as well as of men, and an officer or any other person appointed who is a proper judge, to receive the horses. A plan of Rendezvous should be appointed to which the Recruits of Men and Horses should repair with the riding masters of each Regiment, under the care of experienced officers, in order to exercise and form them for the field. When perfect, to be sent to their respective Regiments. This I think, Sir, would be the most effective and most eligible method of mounting the Cavalry.

MOYLAN COMMANDS ALL THE CAVALRY.

Up to this time there had been five regular cavalry commands —Baylor's, Bland's, Sheldon's, Pulaski's and Moylan's.

March 20, 1778, Washington, on the resignation of Count Pulaski, placed all the cavalry under command of Moylan, as appears by the following order:

As Count Pulaski has left the command of the Horse, never, I believe, to return to any general command in it again, I have to desire that you will repair to Trenton and take upon yourself the command of that Corps, till Congress shall determine further on this head.

You will use your utmost endeavours to have the Cavalry belonging to the four Regiments (now in New Jersey) put in the best order possible, that they may take the field with some degree of eclat. Not a moment's time should be lost in repairing the Saddles and the other accoutrements; and getting the Troopers' Arms completed and repaired. In a word, the Season calls for the utmost vigilance and without descending to the minutia, I shall in general require that you will use your utmost skill and industry to equip in the most economical manner your own Regiment, and cause the others to do the likes.

Inclosed are orders to the commanding Officers of these Regiments to obey you. You must immediately send a relief for the party near the Enemy's lines in Philadelphia County. It is to consist of Men of approved attachment commanded by active, vigilant Officers.

GENERAL PULASKI.

Pulaski "left the command of the Horse" to organize an independent corps, "The Pulaski Legion," in command of which he did good service until his death at the siege of Savannah where, on October 9, 1779, he received a wound from which he died two days later. On January 8, 1855, a monument was dedicated in that city to his memory. History relates that Pulaski was "a devout Roman Catholic," wrote William P. Bowen, Sr., in the Savannah *Morning News* on January 9, 1855.

Count Casimir Pulaski had been appointed Brigadier-General, September 5, 1777.

To Colonels Bland and Baylor the same day the General sent the order:

"Valley Forge, 20th March, 1778: As Count Pulaski will I believe quit the command of the cavalry and is now absent from that corps and at York, you are to receive your orders from Col. Moylan. I expect the officers of your Regiment will remain constantly with the men and use every possible means to train and discipline them."

On March 25, 1778, from Valley Forge, directing that the cavalry be quartered at Iveham [Evesham] and Springfield back of Mount Holly as "very commodious places," so as to get the horses in good condition for the opening of the campaign.

Moylan, it is probable, pastured the horses as Washington advised and thus could operate in that section of New Jersey as opportunity offered and occasion required.

Washington, April 3d, ordered Moylan to send a Corporal and six dragoons to escort our Commissioners to arrange an exchange of prisoners. They must be picked men and horses so as to make the best possible appearance and be very trusty and very intelligent —put the cavalry on the best footing you can.

He advised Moylan of a "certain Mr. Bankson, late of the Continental Marines," whom he suspected as a spy of the British, "though he offered himself as one to us"—"find out his true history"—any thing "that may throw light upon his designs." Manage the business with caution and address.

The Commissioners for the exchange of prisoners met at Germantown the next day—April 4th. The British Commissioners were Colonels Charles O'Hara, Humphrey Stephens and Captain Richard Fitzpatrick. The American Commissioners were Colonel Wm. Grayson, Lieutenant-Colonel Robert H. Harrison, Alexander Hamilton and Commissary Elias Boudinot. The Commissioners met at Newtown, Pa., on April 10th and 11th for conference.

WASHINGTON'S ORDERS.

Washington's orders to Moylan were as follows:

On April 9th. Brigadier-General Count Pulaski is hereby authorized to draught from each Regiment of Horse, two privates of his own choice, with their horses, arms and accoutrements, and one sergeant belonging to Sheldon's Regiment.

VEXATION OF WASHINGTON.

On the 14th Washington wrote the "Commanding Officer" of Sheldon's Regiment expressing his "astonishment and vexation at the low condition of your Horse, which had been permitted to retire to the best quarters offered for the purpose of recruiting them—yet the officers by galloping about the country and by neglect of the horses had reduced them to a worse condition than those which had been kept on duty the whole winter. How you can reconcile this conduct to your feelings as an officer and answer it to your country I know not." [*Ford,* V, 315.]

NEGLIGENT OFFICERS.

On April 11, 1778, from Valley Forge, Washington wrote to Moylan:

Your return of the Cavalry is realy vexatious, but what can be expected when the Officers prefer their own ease and emolument to the good of their Country or to the care and attention which they are in duty bound to pay to the particular Corps they command. In every Service but ours, the Winter is spent in endeavouring to make preparations for the ensuing Campaign.

I desire you will make strict inquiry into the conduct of every Officer present and find out whether those absent have gone upon furlough regularly obtained. And if it appears that they have been negligent in point of duty or are absent without leave, arrest and have them brought to trial; for I am determined to make examples of those to whom this shameful neglect of the Cavalry has been owing. If there has been any deficiency on the part of the Commissary of Forage, let the commanding Officer of Sheldon's make it appear in his own justification.

ARMS AND HORSES.

On April 29th, Washington to Moylan:

I am as much at a loss as you can possibly be how to procure Arms for the Cavalry, there are 107 Carbines in Camp but no Swords or Pistols of any consequence. General Knox informs me that the 1100 Carbines which came into the Eastward and were said to be fit for Horsemen were only a lighter kind of musket. I believe Colonels Baylor and Bland have procured Swords from Hunter's Manufactory in Virginia, but I do not think it will be possible to get a sufficient number of Pistols, except they are imported on purpose. I long ago urged to Congress the necessity of importing a large quantity of Horse Accoutrements from France, but whether the order was ever given or whether they have miscarried in the passage, I do not know.

To this Moylan replied:

MOYLAN'S MOVE ON BORDENTOWN.

TRENTON, 5th May, 1778.

I am exceedingly sorry to find there is so bad a prospect of arming the Cavalry, if the 107 carbines which are in Camp were ordered to this post they would be of some service. I have wrote some time past, to Major Washington, requesting him to procure swords from Hunter's Manufactory, which I hope he will be able to effect.

I have seen but five horses of those purchased by this State. They were sent to Major Clough who rejected them as unfit for the service, I am told there are some tolerable good ones, delivered to the Regiments at Chatham, Brunswick, and Pennington; I propose visiting them this week, and expect from the late accounts I have from each, to make a more favorable report of them than my last, to your Excellency.

Major Clough went yesterday, by my order, with a party of Horse to Bordentown in consequence of the inclosed Letter which I since find was dictated more by the fears of the writers than any real cause for being alarmed. It is true that five or six hundred of the enemy are on this side near Cooper's Ferry covering some woodcutters and hunting for Forage.

I expect the Major back to-morrow and propose ordering that Regiment to Princeton, as Forage is there collecting, and becoming exceeding scarce in this Quarter. Any Commands your Excellency may have for me after this week, you will please to have directed to that place.

MOYLAN REPORTS CAPTURES AND DISASTER.

Moylan to Washington from Trenton, 7th May, 1778:

Major Clough reported that the 63d & 55th Regiments [British] are stationed as guards to a fatigue party of 200 men who are employed cutting wood. The fatigue is daily relieved. Their lines are covered by three small redoubts without cannon. He reconnoitr'd their picket which was strongly posted, sent two of the Militia horse in sight of their lines, which as he expected brought out twelve of the enemies Light Horse, on whom he charged—two of them were wounded, and dismounted, and two others made prisoners not far from their picket, the four prisoners and three horses with their accoutrements are now here and the

Major wishes that the men (who behav'd with the greatest resolution) could soon receive the value of their priezs, as an encouragement to them at the opening of the campaign. The horses are fit for the Regt. and he thinks $100 each will be a moderate price for them. Col. Shreve is at Fostertown in the neighborhood of the enemy, but not strong enough to disturb them, General Dickinson tells me he can draw out four or five thousand men to coöperate in any plan which your Excellency may form. If we could make good a post at Billingsport, it would alarm the enemy exceedingly.

8th May, 1778.

It is nine at night, and am just returned, Sir, from Bordentown, which the enemy left about two hours ago. As near as I could judge there were about 1000 landed there, having previously burnt the Frigates and several ships which were up a creek near that town. I have since seen them rowing up the Pensilvania shore inside of Bules Island. There are three Gallies and thirty-six boats full of British Light Infantry. Mr. Borden's house and two small ones in the town were laid in ashes. It is probable they will come up here this night. General Dickinson has collected about 200 Militia and is in expectation of a larger number coming in to-morrow, he has all the assistance our horse can give him. If they land at this place I will order Bland's Regt. to join us in the morning, at which time I will send this to your Excellency, with any occurrencies that may happen this night.

The 9th. The enemy lay last night on the Pensilvania side about four miles from the Ferry, I am of opinion that they mean to collect the stock & Grain at that side, and not come any farther up the river.

This destructive force of the British was sent from Philadelphia under command of Major Maitland. Twenty or more armed vessels were burned, including Captain John Barry's frigate "The Effingham." Barry at the time was operating in the lower Delaware. He destroyed the forage from Mantua Creek to Port Penn and captured several supply vessels for which he received the thanks of Washington to whom he forwarded supplies at Valley Forge. It was in retaliation for these acts of Barry that Maitland was sent to destroy the vessels of the Americans, which had been sent up the Delaware to White Hill near to Bordentown for security.

Washington replied on May 13th to Moylan's of the 5th:

If the Commissioners of the Navy could have been prevailed upon by me to have scutled and sunk the Frigates last fall, the Enemy would have had little inducement to have visited Bordentown. It would have taken so much labour and time to have raised them that a force might have been sent to interrupt them. Upon the first intimation of the design I detached General Maxwell with a strong party, but the mischief was done and the Enemy gone by the time he reached the Cross Roads. Present my thanks to Major Clough and his small party for their bravery. The price formerly paid the Captors of a Light Horse with his accoutrements was 100 Dollars; but as money has depreciated, the Rule has been in some instances deviated from. Colo. Morgan's Riflemen some time ago took two teams of very capital Horses going into Philadelphia, they were paid 170 Dollars pr. Horse and at that rate, I have estimated those taken by Major Clough. Inclosed you have a warrant for 510 Dollars payable to the Major which he will please to distribute to the party according to their rank.

I do not know whether the Carbines that are here are in proper order. I will have the matter inquired into, and if they are, inform you, that you may send over a Waggon and a small escort for them.

You mentioned in a letter of the 23d April that you understood some members of Congress were dissatisfied with the determination of Rank between Cols. Bland, Baylor, Sheldon and Yourself; if it is so, I have never heard any thing of it from any person but yourself.

Moylan was the Commander of the four Regiments of Horse. Ten days later he notified Washington that if called into action he would not serve as commander of a Regiment but rather as a Volunteer.

MOYLAN SENDS A SPY INTO PHILADELPHIA.

Moylan reported to Washington from Trenton, 13th May:

The late excursion of the enemy prevented my leaving this place, to visit the different Regiments, which I shall put in execution the latter end of this week.

I sent a woman into Philadelphia last Sunday who came out this day, she says that war was certainly declared against France, that General Clinton had taken the command on Monday, General Howe preparing to go home, but is to give a Grand *Fete Champetre* before his departure, that there was great talk of the Troops embarking, the Transports were taking in wood and water, indeed their burning the shipping looks like a move, for if they could hold their present post the ships would consequently fall into their hands. She heard nothing of the Hessians having laid down their arms, which was strongly reported here.

I would not trouble your Excellency with a complaint against the commissary, if I knew who was at the head of that Department, to whom I might address myself, but the great neglect of providing anything (flour excepted) for Bland's and Baylor's Regiments call aloud for redress. They have lived upon salt fish and salt herrings these five weeks past, which is now expended, and nothing have they now but flour to live upon. The issuing commissary says it is the purchaser's fault, but who or where the purchaser is I cannot find out. Mr. Paxton says if he had money he could purchase sufficient. Between them the men are suffering. They have had no pay since October, which with the want of provisions has produced much discontent amongst them. How it is with the other two Regiments I cannot say, but hope they are better off as to provisions. Respecting pay I believe they are in the same situation, which I hope will not continue long, if it does I may dread the consequence, I beg, Sir, that the carbines and what other arms are at Camp for the Cavalry may be sent forward, they are exceedingly wanted.

I have seen General Dickinson who has had some conversation with Mr. Yard, who left Philadelphia yesterday. He confirms what I had from my informant except that there was no positive account of war having been declared but was to have been two days after the last packet left England.

A FORWARD MOVEMENT.

John Laurens, Aide to Washington, on May 17, 1778, wrote Moylan:

His Excellency desires that a select party of fifty dragoons, men that may be depended on, with able Horses, well accoutred and conducted by active partisan Officers, may be ordered to march forthwith to our old camp at Whitemarsh, where they will meet and join a detachment of Infantry, from the Commanding Officer of which they will receive their Orders.

Moylan to Washington:

TRENTON, 23d May, 1778.

I delivered to Major Clough the Warrant for 510 Dollars with your Excellency's thanks to him and his party for their bravery, agreeable to your orders and complied with the orders contained in Col. Lauren's Letter of the 17th as soon as it was possible after the receipt. It reached me at Brunswick the 19th in the afternoon, and as the first and third Regts. were most contiguous to the Delaware, I ordered the detachment from them. I mean to relieve them to-morrow by sending an equal number of the second and fourth, I must continue Major Clough there for some time longer not only because I have an high opinion of his conduct but also that he can be better spared than either of the other two field officers that are with the Brigade. I did expect we should have had more of them, join'd to their Regts. before this time. I have the pleasure to tell you, Sir, that the horses in general are in good order. There are some in each Regiment that never will be, and I am certain it would be saving a good deal of publick money to have them disposed of. The sooner it is done the better.

I have ordered the whole of the horses brought up by Mr. Grey to be delivered over to the third Regiment as they have many men dismounted, there are spare horses sufficient in the 4th to mount the men he brought with him.

My information respecting the Rank of the officers in the Cavalry was hinted to me by General Reed, and the reason for mentioning it was, that should an alteration be made therein to my prejudice, at a time when we may be call'd forth to action, tho I certainly should act as a volunteer, I would not as commanding officer of a Regiment.

Washington to Moylan, Valley Forge, 24th May, 1778:

The Commissary General of Forage has informed me that he can now accommodate the Cavalry in the Neighborhood of the Camp. I therefore desire that you will immediately come over with all the Horse of your own, Bland's and Baylor's Regiments that are in good order; Sheldon's is to remain at Chatham. Good Officers are to be left with the Horses out of condition, who ought rather to attend to getting them in order, than to training them. The weather is growing warm, and it is hardly possible to do both at a time. The three Regiments had better come on in three divisions, at the interval of a day or two, they may then be cantoned with ease, and not be distressed for Forage by coming in in a crowd.

On 28th Washington to Moylan:

As every appearance now indicates a move of the Enemy thro Jersey, I would wish you to continue there until their intentions are more clearly and fully known. If you can subsist the men and Horses at and near Trenton, they will be more conveniently situated there than at any other place, to be ready to observe the Route of the Enemy, and therefore I would have you collect all that are fit for Service as near that place as possible.

General Greene informs me that he apprehends a number of Horses purchased by the agents in this State are unfit for the Dragoon Service, and he would therefore wish to have two or three Officers who are good judges of Horses go round and examine them, that those fit may be sent to the Regiment and the others put to the Draught. Be pleased therefore to send over such Officers and General Greene will direct them where to proceed. I would have you by all means sell those Horses that will never be fit for service again.

Washington's order to "sell all the horses unfit for service" was carried out by Moylan, who by Brigade order dated June 1st, advertised in the *New Jersey Gazette* of the 3d: "That the cast horses belonging to the first, third and fourth Regiments of Light Dragoons be collected at Trenton and sold by publick sale on Monday the 8th inst. at the Market Place."

Moylan to Washington from Trenton, 30th May, 1778:

I had just got as far as White Marsh in compliance with your Excellency's order of the 24th when I received your Letter of the 28th countermanding that order. I have in consequence halted the 3d & 4th Regiments, who were on their march to Camp. The first had proceeded so far, that I thought it best to know your further pleasure before I would give them the order to return. I must beg leave to mention that the arms of these three Regiments are at present in the hands of the detachment with Major Clough, so that it is of little consequence where they are stationed. I will however draw them to this quarter, and depend upon those arms being sent forward which Major Jameson assures me are finished in Virginia & Maryland.

The party under the command of Major Clough consists of

120 horse, if your Excellency thinks proper to keep so large a body upon that duty, I think it will be better to order what remains here fit for duty of the first & third Regts. to join it, and let what are with the Major of the 2d & 4th join their respective Regiments. It will be attended with salutary consequences to have the duty performed Regimentally, both the officers and the men will be better pleased. The former in commanding his own men, the latter in being commanded by their own officers. The Horses by being under the eye of their own officers will be better taken care of. If you approve of it, I will give the orders for putting it into execution.

To which Washington replied June 1st:

I am not a little surprised that the arms of Three Regiments should be reduced to 120. I would wish to have this matter inquired into, and that the Officers of the different Corps send in a return, accounting for so considerable a defection.

As it is probable the Enemy will penetrate the Jerseys I think you had better continue where you are, putting your Cavalry in the best condition for acting which your circumstances will admit of.

Moylan to Washington, Trenton, 2d June, 1778:

Lieutenant-Colonel White informed me that there was a large quantity of flour in a store at Brunswick designed for the prisoners, which lays exposed to the mercy of the enemy. The shalop which was employed to carry it to New York is also full, having been sent back with her cargo. The commanding officer declaring that no intercourse whatsoever will be admitted, until General Winds delivers up a deserter from one of the Jersey Battalions, who came out under the sanction of a flag.

On June 10th Washington ordered Moylan:

" You will immediately assemble all the Cavalry not on duty at some place the most convenient to Camp, where forage is to be had, there to hold themselves in readiness to move on the shortest notice."

On June 18th the British evacuated Philadelphia. Washington moved from Valley Forge to intercept them, which he did ten days later at Monmouth, New Jersey. Captain Heard, of Moylan's Regiment, stationed at Germantown, on June 18th captured a number of the enemy and reported to General Washington:

The in[closed] is a return of prisoner[s] taken by Capt. Heard [ ] as delivered them to the care of Lt. Dover belonging

Capt. McCanes detachment. I have not been able to collect any Matheriall intellegence from them. [*Washington MSS.*, Vol VI.]

Among the captured was "George Sprangle a noted Spie." This was George Spangler. He was executed on August 14th on the Commons, now the site of the City Hall, for "acting as spy and guide to the British army."

Moylan's reports to Washington, Trenton, 23d June, 1778:

General Reed was down with me in view of the enemy. He can therefore inform you of every thing material, I have ordered Colonel White with a Squadron of Horse into the rear of the enemy, whose van I believe to be at this time in Allentown. He will keep me constantly advised of what passes in the rear, and the remainder of the Horse will be engaged on their front & left flank. You may depend on having the earliest inteligence of their motions, that I can from my own observations & of the officers under me collect.

To General Lee, Moylan reported from on the Middletown road five miles from the courthouse:

The enemy turned off at Carnans fields to the Shrewsburg road. There is a place called the falls about 6 miles from Carnans, where there is another road to Middletown. Should they choose to go that route I believe they will halt at the falls, as it is advantageous ground. If they do you will hear soon from me.

To "Baron Steuben or any General officer in the American Service," Moylan reported from Toponamos Church, 27th June, 1778:

A rascal who was trumpet Major to the first Regimt. deserted to the enemy last evening which obliged me to move off my party from Longstreets on the hill from whence the Baron wrote yesterday. I left an officer and four of the best mounted, who were obligd to retire from a large party of the enemies light horse at day break this morng. I met them here, and they are of opinion that the enemy were then in motion, I have sent a party to the same place, and another to the Shrewsburg and Middletown roads from whom I expect to hear in an hour. I am just going upon the same rout, with horses almost wore down. I wish some of White's party were sent to me otherwise I shall not be able to do anything satisfactory. When I can be certain of this rout I will dispatch a horseman with the inteligence.

To Washington he reported from Longstreets, 27th June, 1778 [11 A.M.]:

Every thing looks in the same situation as yesterday; at Freehold, we took three prisoners whom I send to your Excellency, they say it is the opinion which prevails in their Camp, that they will march to-morrow morning for So. Amboy, I saw a man yesterday from Middletown point, who says there were no vessels of any consequence in the Bay, that there was no collection of boats at Princes Bay. I expect an officer in soon from the Shrewsburg road. allso the great Midleton, there is one road to the later place which passes near to the post I now write from.

To Washington from Scotch Meetinghouse, 27th June, 1778:

I am just come from Longstreet's hill. I sent an officer since sunset close in to the lines and from his report, with my own observation there was no appearance of a movement, I sent an officer on the Middleton & Shrewsburg roads, who makes the same report. I will be out before the sun, & if anything new you will hear from me.

At "½ past two, Sunday," Moylan reported:

At twelve o'clock the enemy were halted at Polhemos hill which is on what is calld the fifteen hundred acres. They are now again in motion & seem to bend their course towards Middletown, thoro bye woods which were not suspected to be passable, but there are so many intersections in the roads that it is impossible to judge whether they will go to Middletown or go on to the falls, I have them full in view, and must move as a party is endeavoring to surround me.

Within an hour the Battle of Monmouth was being fought and a victory won by Washington. The treachery of General Charles Lee came near making the day one of disaster for the Americans. He was Court-Martialed and suspended. It is now believed he was a traitor for having arranged the plan for the capture of Philadelphia while a prisoner in New York.

James McHenry, Secretary to Washington, wrote to his friend, George Lux of Baltimore, that after the Battle of Monmouth, "Colonel Moylan's dragoons are still hanging on the enemy and waiting to see them safely a-ship board." [*Mag. Am. History,* III, 357.]

Moylan to Washington, 29th June, 1778:

Within three miles of Middletown we attacked a party of the

enemy this morning and took one Captain, one Lieut. and one Ensign with two privates, prisoners, & killed a few more. The British Army is expected at that place this day or to-morrow; quarters are taken for Gen. Clinton at Midletown, and for Lord Cornwallis at the place where we made the attack, which I suppose will be the rear of their army. The baggage is still where it last night halted, badly guarded. I wish there was infantry in this Quarter, a great stroke might still be made upon it.

An hour later he again reported:

Captain Plunket will deliver to your Excellency the prisoners mentioned in the note I sent you about one hour ago, I since find that General Clinton's Qrs. are taken at the place where the prisoners were taken. I am informed that there were fifty horses taken in the engagement yesterday, I wish it may be true, as I could immediately bring as many men into the field with good fresh horses provided I can [get] the arms and accoutrements.

He reported on 2d July, 1778:

I have sent orders to the different parties that are now out to come in this evening; they ought to have at least a fortnight's rest before they begin to march and if your Excellency would approve of it I would recommend Shrewsburg and its environs for that purpose. It is inhabited by the disaffected who as I am informd have large quantities of grain and the pasture there is exceeding fine. The enemy are now four miles from Midletown, I expect they will be embarked to-morrow or next day; Morgan's & Gist's men with the parties of Horse, have saved a fine country from being pilaged.

Colonel Daniel Morgan writing to Washington, 2d July, 1778, from Middletown, N. J., said:

"I am and have been ever since I came out, at a great loss for light horse, having none with me. Gen. Scott sent me a Sergeant and six whose horses were tired and rather an incumbrance as they could scarcely raise a gallop. Major Jameson was here yesterday. I applied to him for a few; he sent Captain Harrison who staid with me about 2 hours, when Colonel Moylan sent for him and his party. Colonel Moylan certainly has reasons for so doing, but, Sir, you know the cavalry are the eyes of the infantry, and without any, my situation must not be very pleasing, being in full view of the enemy's whole army." [*Corres. Rev.*, II, p. 153.]

On July 5, 1778, Washington's order by Colonel Clement

Biddle from Camp at New Brunswick to Moylan at Bound Brook was: "If the cavalry should halt to refresh themselves I am of opinion they can be best furnished with hay and pasture on the plain below the Mountain from Middlebrook to the Scotch Plains and I now write to Mr. Furman to use his utmost endeavors to procure grain suitable for them. Some corn is already ordered from Trenton which Mr. Furnan shall detain for you. You will please to inform him the route you will take when you march or rather before that he may lodge the necessary forage at proper places."

On the 7th Washington ordered:

"That you collect the *whole of the cavalry* without delay, as well the unarmed as the armed, and after a little refreshment, and getting the horses shod, &c., proceed moderately towards the North River to join the Army.

Major John Jameson writing Colonel Bland from Bound Brook, New Jersey, July 9, 1778, after the Battle of Monmouth, said: "Our men are so naked that it is a shame to bring them into the field; pray send some officer with clothes for the poor fellows. . . . Colonel Moylan has appointed a Brigade Major as commandant of the cavalry." [*Bland Papers,* I, 97.]

Washington, replying to a later letter of Bland's, said:

"The officers who had the care of procuring necessaries for Moylan and Sheldon's Regiments have long since completed the business and the men are well equipped." [*Ibid,* 98.]

On the 16th July Washington's order from Haverstraw was:

"That you proceed immediately with the horse under your command to Orange Town where you will find Capt. Hopkins who has instructions for the Cavalry."

MOYLAN AND MISS VAN HORN.

To this Moylan replied:

TAPAN, 23d July, 1778.

Your orders of the 16th reached me the 19th instant, and agreeable thereto I have marchd the three Regiments of horse to this place.

I have seen your Excellency's instructions to Captain Hopkins to which I will pay due attention. The English neighborhood, would be a good place for the Cavalry. If they are to stay any time on this side the river, I shall expect further orders from your Excellency by the bearer.

P.S.—Am engag'd to Miss Vanhorn.

Moylan's postscript shows that, amidst war's alarms and desolation, Cupid's arrows struck two hearts with but a single thought. "Miss Vanhorn" was one of the five daughters of Philip Van Horne, a Colonial Colonel of New Jersey Militia.

Moylan, "the fascination of whose merry nature and fine appearance, the latter enhanced by his red waist coat, buckskin breeches, bright green coat and bear skin hat, were too great for the Middlebrook beauty to withstand." [Mellick's *Story of Old Farm,* p. 480.]

In consequence of the army wintering at Middlebrook the five daughters of the Colonel of the Colonial Militia found husbands. [*Mag. Am. His.,* 1890, 153.]

But Colonel Moylan had other engagements than that with Miss Van Horne of which he had informed General Washington who in his reply tendered no "congratulations." He was occupied with more important matters than those of his officers' matrimonial engagements.

He wrote Moylan from White Plains, 25th July, 1778:

I think the best position for the Cavalry to answer the purpose of foraging and covering the country will be about Hackensack New Bridge. You then have an opportunity of drawing supplies from the Country between the North River and Hackensack, and Hackensack and Pasaic as your station will be central. You will therefore be ready to move at a moment's warning.

To which Moylan made reply:

ORANGE TOWN, 26th July, 1778.

I propose to move what horse are fit for duty to the New bridge to-morrow agreeable to your orders and shall endeavor to fulfill the duty recommended therein.

Whenever your Excellency will order us over you will find the Cavalry ready to obey your commands.

On the 29th Moylan, from "Hackensac," wrote Washington that he had "come with the Cavalry to this neighborhood. On my arrival I reconoitred the country and found a great majority disaffected and taking every opportunity of supplying the enemy.

Yesterday I sent a party of 80 horse to Bergen with orders to drive up what cattle they could collect from that town, to the point, which they have effected by bringing with them near 300 head of horned cattle, 60 sheep, some horses, mares & colts. Many of the first are milch cows, and tho its certain that the milch & butter is for the chief part sent to New York from that Quarter, there appears a great degree of cruelty in taking from a number of famillys, perhaps their only support. I am teased by the women. and with difficulty can prevail on my feelings, to suspend my giving to them their cows, untill I have your Excellency's opinion and orders on the subject—this manavre has alarmd the City, Powles Hook & the encampment on Staten Island. The Fort was manned. So was the Redout at Powleshook, and the army at Staten Island turned out, to the amount, as near as could be judged by Major Clough (who commanded the party) of 3000, tho their encampment would promise 5000.

"I have just come in from Fort Lee. The heights from Harlem up to Kingsbridge are interspersd with Tents, the chief encampment on York Island seems to me to be at Fort Washington. Those immediately about the Fort are Hessians. There is a pretty large encampment on your side of Spiten Devil Creek—and a redout with a magazine in its center—one ship pretty near on a line with Col. Morris' horse, another with three small craft near the entrance of the above mentioned creek, are all the vessels in the North River that I could discover at 12 o'clock this day. A report prevails of a French & Spanish fleet being at the Hook. It is believed at Bergen, which your Excellency knows is but four miles from New York."

To which Washington replied from White Plains, 30th July:

I approve of the step you took to drive off the Stock from Bergen, but if it appears to you that the families will be distressed by keeping their milch cattle, you have liberty to restore them to such persons and in such numbers as you think proper.

I desire you will come over with all the Cavalry except about twenty-four, who are to act in concert with the detachment of foot. If that number appears too few, you may increase it to any as far as fifty. Colo. Simcoe told Capt. Sargent (who went down with a flag yesterday) that Admiral Byron was Arrived. Be pleased to endeavour to find out the truth of this.

MOYLAN THE RANKING COLONEL.

The annexed document shows that Colonel Moylan was the ranking officer of the Dragoons:

Thus the dispute as to relative rank was settled by the Field Officers as Washington suggested to Moylan, April 11, 1778.

An Extract from General Orders of August 7, 1778:

The Rank of the Field Officers of the four Regiments of light Dragoons having been settled by a Board of Genl. Officers at White March on the 24th of November last. The Officers are to Rank in the following Manner agreeable to the Report of the Board:

	Cols.	Lieut.-Cols.	Majors
1.	Moylan	White	Washington
2.	Baylor	Byrd	Jameson
3.	Bland	Temple	Clough
4.	Sheldon	Blackden	Talmage

The above is a true Copy of the genl. Orders of August 7, 1778.

S. MOYLAN, Col. C. L. D.

[*Letters of Washington,* 152, Vol. VI, p. 379.]

MARRIAGE OF MOYLAN.

The *Pennsylvania Post* of October 7, 1778, announced in letter from "Trenton, September 30th":

"On Saturday, the twelfth inst., was married at Phill's Hill by the Rev. Mr. Beach, Stephen Moylan, Esq., Col. Commandant of the American Light Dragoons, to Miss Mary Ricketts Van Horne, eldest daughter of Col. Van Horne; a lady possessed of every accomplishment to render the married state happy."

Colonel Van Horne lived at Phil's Hill on Middlebrook stream a mile west of Bound Brook. He had been Colonel of a New Jersey Militia Regiment under the English government. Washington regarded him as a suspicious character. On January 1, 1777, Washington wrote to Colonel Reed: "I wish you had brought Van Horne off with you, for from his noted character there is no dependence to be placed on his parole."

On January 19th he wrote: "Wouldn't it be best to order P. Van Horne to Brunswick? These people, in my opinion, can do less injury there than anywhere else."

Or was this another Van Horne?

Captain Graydon in his *Memoirs* relates that on his release from captivity in New York he stopped at Van Horne's who, notwithstanding Washington's desire to have him at Brunswick, "kept his post at Bound Brook where he alternately entertained the officers of both armies being visited by one and sometimes the other. His hospitality ought certainly to have been recompensed by an unlimited credit on the public stores. His house, used as a hotel, seemed constantly full. Notwithstanding the number of guests that were provided for, there appeared no deficiency in accommodation and we supped and lodged well."

Simcoe's Queen's Rangers in October, 1778, went to Bound Brook intending to capture Moylan at Van Horne's but he was not there. [*Journal*, p. 114.]

The children of Colonel Stephen Moylan were a still-born child, March, 1780, at Middletown, Connecticut; two daughters, Maria and Elizabeth Catharine and an infant interred in St. Mary's graveyard, Philadelphia, February 24, 1795.

Maria was baptized at St. Mary's, Philadelphia, March 5, 1786. Thomas FitzSimons and wife and John and Jasper Moylan and Isabella Masse [or Wasse] were sponsors. She married Samuel Fox. They had children, viz: Elizabeth Moylan Fox, b. 1806, died unmarried. Mary Moylan Fox, b. 1808; m., 1832, Henry D. Bird, of Virginia. Stephen Moylan Fox, b. 1810; m., 1838, Louisa Linton. Margaret Fox, b. 1812; m., 1839, Dr. Tarleton B. Amberson. Anna Fox, b. 1815, d. 1816. Philip Lansdale Fox, b. 1817; m., 1845, Elizabeth DePui. Violetta Lansdale Fox, b. 1819; m., 1842, David M. Courtney. Edward Fox, b. 1821, d. 1862; m., about 1844, Sophia (?).

Moylan's daughter Elizabeth Catharine was married on March 10, 1807, by Rt. Rev. William White, Episcopal Bishop of Pennsylvania, to William Moylan Lansdale, her first cousin. His father, Major Thomas L. Lansdale, had married Cornelia Van Horne.

They had four children: Philip, Medical Director U. S. N. Maria married John W. Hornor; Cornelia, married Maskell Ewing, of Havre de Grace, Maryland. She died in Philadelphia January 31st, 1906, aged 86 years, leaving surviving her two sons and two daughters: J. Hunter Ewing, of the banking house of Townsend, Whelan & Co.; Maskell Ewing of the firm of Ewing & Longacre; Mrs. Louisa B. Gallatin of New York and Miss Cornelia L. Ewing of Philadelphia. Caroline married Edmund B. DuVal. All of these left descendants, but none of General Moylan's children or later descendants professed his Faith. His daughter Elizabeth Catharine's four children had seventeen children.

On September 12, 1778, the marriage day of Colonel Moylan, General Washington was at White Plains, New York, where he had been encamped from July 24th.

Nothing appears on records at hand to show the newly married Colonel's military operations until Colonel Richard H. Meade, one of Washington's Aides, writing from Fredericksburg, N. J., on November 2, 1778, to Moylan, said:

The bearer, Mr. Simeon Newel, is the Gentleman who contracted with Capt. Hopkins to supply many different Articles for the use of your Regiment. He has found some difficulty in the settlement of his accounts, to remedy which he has laid the matter before His Excellency, by whom I am directed to request that you will take the necessary steps in order that the matter may be accomplished according to right. Mr. Newel seems to have taken every precaution, and to have complied with Capt. Hopkins's directions. It will be proper to call on the Capt. when it is hoped the matter will be settled without much difficulty.

"THE HAPPY EVENT."

General Stirling, October 8, 1778, from New Brunswick, to Colonel Moylan at Phill's Hill:

"I have just seen Col. Van Horn's letter of yesterday to General Maxwell enquiring the situation of the enemy in this State. They are at present stretched from the New Bridge above Hackensack across towards Fort Lee with two redoubts on the heights on this side the bridge; their strength is about 7000 of their best British troops; on the 5th they advanced to within about three miles of this place but in the evening retired to their present situation. I believe they have nearly completed their forage and will soon quit this State.

"I most sincerely congratulate you on the late happy event of your new connection with the most amiable of Ladies. I beg to present my sincere respects and best wishes for your mutual happiness and my best compliments to all the family."

Moylan's command was quartered at Lancaster, Pa., when Washington sent this order:

ORDERS TO MARCH TO CONNECTICUT.

You will forthwith proceed to Durham, between New Haven and Hartford in the State of Connecticut, with your Regiment of Cavalry, where you are to fix your quarters for the Winter.

In quartering the Regiment at Durham you will preserve as much compactness as the nature of the place will admit, that by having them all under your own Eye, you may be able to keep up good discipline, and prevent dissipation and irregularity.

It is not designed that the Regiment should do ordinary duty, or be called out upon every common occasion. But in case if invasion, or the advance of the Enemy, you are to obey the orders of General Putnam, and assist in giving them every opposition.

BRITISH ARMY REPORTS MADE KNOWN TO WASHINGTON.

It was, probably, while at Durham that the "circumstance" herewith related took place:

Jones' *History of New York during the Revolution,* Vol. II, p. 210, relates "a circumstance told me by a gentleman who was a prisoner in Connecticut during the winter of 1779. Stephen Moylan Esq., who commanded a Regiment of horse in the rebel service and was quartered in the same town, in an evening's conversation told the gentleman that not a return of the number and state of the British army at New York had been made to General Clinton for the last two years, but that General Washington received a copy of it in twenty-four, or at most, in forty-eight hours after its delivery to the Commander-in-Chief."

At Durham Moylan's Dragoons were quartered all winter and the spring of 1779. General Washington, on June 28th, sent from New Windsor these instructions:

MOYLAN'S COMMAND INCREASED.

When you have crossed the North River with your Regiment you will proceed to the neighborhood of Bedford [N. Y.], where Col. Sheldon's Horse and a few light infantry are stationed, these you will take under your command. The purpose of this command are to protect the country and inhabitants, give confidence to the militia, and as far as it lies in your power, gain intelligence of the Enemy's force, movements and designs of which you will give me the most punctual information.

I leave it to your own Judgment, from an examination of the

country, and according to the circumstances, to take a position that will best answer these purposes, consistent with the security and accommodation of your Troops.

Col. Armand's Corps I propose ordering down, who will also be under your command.

The uniform worn in this campaign is described: Short green jackets, red waistcoats ["the green above the red"], buckskin breeches and leather caps trimmed with bear skin. [*Am. His. Reg.*, IV, 502.]

Washington, from New Windsor, New York, on July 10, 1779. notified General Heath, "the enemy in considerable force was moving by land towards Horse Neck, with artillery and wagons, and a detachment sent out from New Haven gone to Fairfield, burnt the town, reimbarked and were off Norwalk where tis imagined they will land and destroy that place and the two bodies join to ravage and distress the country. You will march to-morrow morning as early as possible. You will direct Colonel Moylan with the Cavalry and infantry under his command to join you at such place as you may think proper." [*Mass. His. Soc. Col.,* 5 S., IV, 109.]

Washington wrote Moylan from New Windsor, July 10. 1779:

The Person you mention is employed by me. I place a good deal of confidence in him. though he is obliged in order to answer our purposes, to appear friendly to the enemy. I thank you for the intelligence you communicate. The ravages of the enemy particularly at this Season, are distressing; but our situation makes it impossible to prevent them. Armand's Corps has been directed to join you.

DESTRUCTION OF NORWALK.

Colonel Moylan to Washington from Wilton, 12th July, 1779:

The day on which I last addressed myself to your Excellency General Parsons ordered the Infantry which were under my command at Bedford, to march to Norwalk; finding my small party of Horse would be of little consequence in that quarter I march'd them also to Norwalk where I arrived yesterday morning just at the time the enemy had made their landing good—an engagement very soon commenced, and a vast deal of ammunition wasted, to very little purpose. as in general our militia kept at awfull distance. The few men, 150 in number. of the Infantry behaved exceeding well, maintaind their post with the greatest bravery. They were deserted by the militia and order'd to withdraw, which they did without the least appearance of confusion. The town was shortly in ashes and now remains a monument of British barbarity—they reimbarked under cover of the smoke, which was right in our faces and I suppose Stamford will next feel the cruel effects of their rage. It was a fortunate circumstance for me that I kept my Qrs. on Saturday evening, for four hundred of the enemies horse came up to Bedford the next morning and as I am informd burnt it. They did not proceed further as they probably were informd of my departure.

I order'd Colonel White who commanded the Horse which were watching the army near Rye to join me at this place which he effected last evening. Mr. Gill a Lt. in my Regiment took four of the enemy prisoners, one is so badly wounded that I believe he cannot recover. He had some conversation with Sir Harry Clinton without knowing who he was, until told by the prisoners one of whom I should inform your Excelency is of that Corps who receive no pay, but are supported by plundering the inhabitants. I wish he had been cut down, but I believe he will meet with his deserts, as I intend to send him to Governor Clinton to whom I find he is well known. Both Regiments are now very much fatigued. I will keep as many with me as are best able to bear the hardships of this rough country and send the remainder to some place where they can rest in some kind of security, which I assure you Sir is difficult in any place near Bedford without a body of Infantry to support them. The enemy being so vastly superior to us in Cavalry. My reason for staying here is to give countenance to the militia, who seem to place some confidence in me by doing my duty yesterday. I am very confident that General Clinton's movement to Rye and that Quarter was intended to draw your army or part thereof away from the Fort which I have little doubt is the main object with him of this Campaign. A Letter from General Heath mentioning his advance towards Bedford with a Division of the Army. He orders me to move forward with the Cavalry and Infantry under my Command to join him at Hoits. I am sorry to quit this neighborhood where I know I should be of some service in keeping the militia together, but his orders shall be obeyed.

Colonel Moylan sent this report to Washington from Ridge Field. July 21, 1779:

I am again detached from the left wing of the army, having left 40 horses at Peekskill and its neighborhood under the command of Capt. Hopkins in order to watch the enemies motions, and procure inteligence from that Quarter. As I am without Infantry I cannot venture lower down, with the horse. Indeed the marching and countermarching we have had since I left Norwalk has left but few fit for duty. After they have a little rest I propose sending out small parties to gain inteligence, which is all that can be expected from our numbers compared with the enemies Cavalry. I sent a flag in last Saturday by an inteligent officer, who I had not seen until this morning. He assures me that the main body of the enemies Infantry had not marched on Sunday as he could plainly discover that they lay at the west side of the bridge leading in to East Chester, about fifteen hundred with five hundred horse were marchd to Newbridge over Groton. These Troops came to raise the siege of Verplank's point and considering the besiegers situation it was forunate they did not push their march. There is no acct. from Armand's Corps. General Heath promised to send me 150 Infantry. If it could be made up 200, it would be little enough for the Duty they will have to go thoro'. If your Excellency thinks it proper to order them on I shall be enabled to move on to Bedford as soon as they may arrive.

At this time the enemy were "advancing into Connecticut. No troops but the militia there to oppose them," wrote General Parsons to General Glover, 10th July, 1779. Grover moved his forces and arrived at Ridgefield on 22d where on 23d he received orders from Washington to "halt until further orders." Moylan was ordered to join him. Washington was then at West Point, New York. On the same day (24th) he ordered General Heath, at Mandeville, "to order Colonel Moylan to collect his horse and join General Glover under whose command he will be for the present." [*Mass.*, 5, IV, 114.]

General Glover had left Providence on July 7th to join Washington. On 16th Washington ordered Heath, at Canaan, to "March to-morrow morning towards Peekskill," where he would find General Howe with a couple of brigades—"the command of the whole will devolve on you. Send to hasten on Glover's brigade to join you at the same place." Heath was to follow Howe's instructions which were to open a battery on the enemy at Verplank's Point—which were to be one of the four brigades should be stationed in the gorge of the mountains and the other three proceed to the highlands opposite West Point. On July 19th Heath was ordered to direct Glover to halt with his brigade at Ridgefield until he is further instructed.

On July 30th Heath reported to Washington an action of Captain Hopkins of Moylan's Cavalry which caused Washington, the same day, to write Heath: "Capt. Hopkins' conduct really deserves applause and shows spirit of interprise that does Him honor."

On August 1, 1779, Washington, then at West Point, New York, wrote Moylan:

General Howe has gone to Ridgefield, to take the command of Glover's Brigade, and all the Troops in that Quarter, and will make such dispositions of them as may appear best. I am sorry it is not in my power to send you any hard money. I have but little and it is more particularly intended for Persons within the enemy's lines. If you will make out a return of your dismounted Men, and the necessary Arms and accoutrements, I will send for them. I will direct them to be supplied.

On the 6th he wrote he had received Moylan's of the 3d "and am persuaded you had made a good disposition of the Troops under your command. I believe there has been no embarkation of the Enemy, except for the Marines, on board the Ships said to be gone in pursuit of our armed Vessels, on the expedition to Penobscot."

Mrs. Moylan was ill at her father's at Bound Brook, New Jersey. Moylan applied to Washington for furlough and received this reply:

SECRET SERVICE MONEY.

Mrs. Moylan's illness will readily obtain my consent to your being absent from the army a fortnight provided a movement of the enemy should not require your presence sooner. General Howe should be made acquainted with your absence. The Sum you speak of, as having expended for secret Service, surprises me exceedingly, because I do not call to mind ever having empowered you to lay out Money for such purposes nor do I recollect ever to have received any intelligence, of an extraordinary nature, differing in any respect from that which every Officer at an advanced Post, or removed from the main army, regularly obtained (by his own observation and industry, or from the Inhabitants), transmitted to Head Quarters, and because the Sum exceeds the aggregate of the charges of all the other Officers of the line, for Services of this kind, although some of them have been appointed for, and have attended particularly to this business. Under these circumstances and as a public Officer, my duty obliges me to call for such an account, as will justify my conduct, in ordering payment.

MOYLAN'S WORD HIS VOUCHER.

To this Moylan replied:

GREENWICH, 13th August, 1779.

Accept my most grateful thanks for your kind permission to pay a visit to Mrs. Moylan. I will, you may be assured Sir not lose time from my duty, which I assure you no officer in the army is fonder of doing than I am. As to the expence of Intelligence I give you my word and my honor I have been rather under, than over, in what I mentioned to your Excellency in my last. This, or rather Poudridge post, has been the most expensive as I promised a man who had been four times within Kingsbridge to make his Dollars Silver, and he says and I believe the last bills I gave him, which were two, one of 65 & one of 45 Dollars, he got but at the rate of 12 for one—as General Howe is at present in this place and will take every opportunity of gaining intelligence, I have not occasion to interfere in that department. At the same time I must remark to your Excellency that I had positively your orders, when I [was] with you at Middlebrook, to lay out money to gain intelligence, and when I had the honor to receive your commands of taking the command of the cavalry in the Jersey in 1778 I asked you whether I should try to gain inteligence. Your answer was "yes by all means," which made me fix my Qrs. at Trenton. Indeed the expence for information at that post was very trifleing, but if your Excellency will reccollect what I sent you from Amboy and in that neighborhood, you will see that it must have come from persons I employed within the City. I mean persons I sent in—to New York—the countermarch of the army from the Clove in that year I realy thought was occasiond by the information I had given to your Excellency.

If my word is not a sufficient voucher to the public. I assure your Excellency I will not nor cannot give any other, and if I even do get what I laid out I do not think from the depreciation of the money that I shall be paid half what in justice I am entitled to. The freedom with which I write to your Excellency I dare say from my knowledge of you, you will pardon, for you may be assured it does not arise from presumption. I know your heart, it is a great, a good one, and amongst your admirers there is no one who can subscribe himself with more propriety, your assured friend and affectionate Humble Sert. than

STEPHEN MOYLAN.

General Arthur St. Clair, writing to President Reed of Pennsylvania from West Point, August 24, 1779, said:

In Conversation with Col. Moylan yesterday, his Regiment came upon the Cappet. A Resolve of Congress seems to have had in view that the Regiments of Horse that have been raised in particular States should be considered as part of their Quota in the Continental Army. He would be very happy to find himself in that situation; and tho' the officers have not all been taken from Pennsylvania the Men were, I believe, all raised there. I will be obliged to you if you will please to communicate your Sentiments on that head.

RE-ENLISTED MEN TO BE NATIVES.

Washington to Moylan from West Point, 24th September, 1779:

The return of Clothing necessary for your regiment, should be made, by you, without delay, to the Board of War, who will give you information where, to whom you are to apply.

If any of your present men, whose time of service will shortly expire, will re-inlist for the war, they will be entitled to a bounty of 200 dollars and ten dollars to the Officer re-inlisting them. If you find any in the above predicament, willing to re-inlist, you can send over for a sum of money for that purpose. Colonel Sheldon had liberty to endeavour to inlist as many new recruits, as would compleat his regiment to the establishment, provided they could be obtained upon the terms of serving with the regiment, as dismounted Dragoons, until there should be a necessity or conveniency of mounting them. This to be clearly expressed in the inlistment, that the men may have no pretext for complaint on being made to serve on foot. They were also to be inlisted for the War, and no temporary engagements entered into, on any account whatever. The bounty to new recruits to be 200 dollars and 20 dollars to the recruiting Officer. These men must be natives, of good character, and every way suitable for dragoons. If you are of opinion that you can obtain men upon the foregoing terms, you may draw money and try the experiment.

On September 30, 1779, John Pierce, Deputy Paymaster-General, "paid Patrick Bennett for recruiting the 4th Regiment of Light Dragoons to be accounted for by Colonel Moylan, $5,000." Washington the next day wrote Moylan from West Point:

"I have given Mr. Bennet a Warrant for 5000 dollars for recruiting, which sum he will deliver to you. The state of the military chest will not allow of a larger sum at this time, but you may have more when this is expended."

Moylan to Washington from North Castle, November 5, 1779:

I have received Coats and Waistcoats for the 4th Regiment, and am just now informed by Captain Hunter of Bedford that he has as many pairs of Leather breeches of the best quality as will cloath the Regiment which he will part with, if your Excelly. will be pleased to give me an order. I will purchass them from him on the best terms I can—if I have your approbation for procureing the breeches and an order for shirts, stockings & boots—the men will be enabled to keep the field and I am convinced, will do as much duty if not more, than any equal number of men in the service.

On November 20, 1779, Washington wrote Governor Trumbull of Connecticut that he would "station New Hampshire troops at Danbury and Moylan and Sheldon's Regiments to east of that." [*Mass. His. Soc. Col.,* X, 148.]

Washington to Moylan from West Point, 27th November, 1779: "You will find in the inclosed instructions the place for the cantonment of the Cavalry and the limitations for furloughing both Officers and Men."

The Instructions read:

INSTRUCTIONS.

As soon as the division under the command of Major-General Howe moves to its ground for Winter Quarters, you will proceed with your own and Sheldon's Regiment of Dragoons, to such place or places as the Quarter Master General may have assigned you for Winter Quarters. This may be in Wallingford, Durham or Hadham, as conveniences and Forage may best suit, or in case of necessity you may remove the whole or part to Colchester. In the cantonment of the regiment, you will preserve as much compactness as the nature of the place will admit, that by having them all under your own eye, you may be able to keep up good discipline, and prevent dissipation and irregularity.

It is not designed that the regiment should do ordinary duty, or be called out upon every common occasion. But in case of invasion, or the advances of the enemy you are to obey the Orders of General Poor, or other your superior officer commanding at this post and assist in giving them every opposition.

Washington moved into winter quarters at Morristown, New Jersey, where Moylan, on December 15th and 16th, wrote him relative to quartering the cavalry. Mr. Hubbard declaring the location unsuitable, and Moylan that it would well provide for the wants. However, on December 20th, Washington informed Moylan:

On a representation of Mr. Hubbard, that the difficulty of obtaining forage and other supplies for the two regiments of Dragoons at Middletown and Weathersfield, would be very difficult and productive of an enormous expense, I am to desire you will remove them to Colchester, where a Magazine of Forage is laid in, and a sufficient quantity of Stable room can be provided.

Washington, replying to Moylan's of 15th and 16th from Morristown on Christmas Day, 1779, said:

I am extremely sorry that the question of quartering the cavalry stands upon so very disagreeable a footing between Mr. Hubbard and yourself. But there are reasons which will not suffer me to retract the order contained in mine of the 20th. Though I doubt not the cavalry may be well provided and accommodated at the places you wish, yet I prefer Colchester, because large Magazines are already formed there and other preparations made. I am told too, all your Wood, where you now are, must be brought several miles at an enormous expense; at any rate fresh purchases of forage must be made, which in the present exhausted state of the treasury is scarcely practicable; or if practicable, unadvisable. You will therefore remove to Colchester.

I wish to receive from you, by the earliest opportunity, a Return of the Officers and Men in your Regiment who belong to the State of Virginia. You will mention the names and rank of the first, the number of the latter will do, in which you will note how many of them engaged for the War and to what other periods the rest stand engaged.

"NOT MY BUSINESS TO DISPUTE."

To which Moylan replied from Middletown, 4th January, 1780:

I have ordered Sheldon's Regiment on from Weathersfield to Colchester. The 4th is at Walingsford where they will remain a few days in order to give Mr. Hubbard time to make some preparation for their reception which you will see by the inclosed report is necessary—the representation made to you, Sir, were not founded on facts, but it is not my business to dispute, but to obey your orders which I shall put in execution as soon as possible—you will receive by this conveyance returns of the two Regiments for last month, and you may be assured of my transmitting them monthly agreeable to your orders. You have also a return of the officers and men belonging to the 4th Regiment who are from the State of Virginia. The weather here is very severe, and many, indeed the Majority of the Dragoons, have neither boots or shoes. I have spoke to Capt. Starr of this place who promises me to supply them with shoes. As it is a case of necessity I hope neither he or I can receive censure.

Moylan's "business" was "not to dispute" but it turned out by later events that he was justified in not approving of the selection of Colchester. The Governor and Assembly of Connecticut were likewise opposed to the cantonment of the troops there. Mr. Hubbard also confessed the selection unadvisable.

Washington wrote Moylan from Morristown, 5th January 1780:

The Board of War are anxious to compleat an arrangement of the four Regiments of Cavalry and have wrote to me on that account. You will therefore be pleased to forward that of your Regiment as speedily as possible.

To this Moylan replied:

MOYLAN'S "A SOUTHERN REGIMENT."

MIDDLETOWN, 17th January, 1780.

The exceeding heavy snows had stoppd up the roads in such a manner as rendered it absolutely impossible for the 2d Regt. to move on to Colchester or the 4th to leave Walingford, until the 14th inst. when they got as far as Durham, I expect them in from thence at this place to-morrow, where they shall remain no longer than untill a road can be found passable to Colchester, as the weather is cleared up, I hope they will be able to march in a day or two. The river will now bear, and I am informed there is a road by marching 16 miles on it that will probably be beat sufficient. I have sent down to have it examind, and hope for a favorable report. The Inhabitants of the town of Durham, instigated by Mr. Wadsworth formerly a Bridgr. Genl. would scarcely let the Regiment halt at that place, tho in their direct rout to Colchester, for no other reason that I know of, but that they are a Southern Regiment, which I am sorry to say, is not a reccommendation in this State. I find this Gentn., if I may call him one, has represented me in an unfavorable light to the Governor, from whom I have this day received a most insulting Letter. As I know it was wrote in prejudice I will not give such an answer, as it deserves. I have one pleasure which is that no Regiment could be more orderly than the 4th since they have come into this State, and I have no doubt but they will continue so.

A further report was made by Moylan from Middletown, 22d January, 1780:

There is at last a path made from East Haddam to Colchester, by which rout I shall march off the Regiment this day. We have an exceeding cold day, and the Regiment so very badly off for undercloaths that they are much to be pitied. If the Quarters are so bad as represented to me, it will be much to be lamented that the whole has been ordered thither—Major Tallmadge informs me that a part of the 2d will march this by way of Bolton—and the remainder the 24th—the dismounted of that Regiment are to remain at Weathersfield, as it will be impossible to quarter them at Colchester. This is done at the request of Mr. Hubbard, who acknowledges that the men cannot be quartered any where convenient to the stables.

Captain Pike & Captain Craig are going to recruit for the Regiment, if your Excellency will be pleasd to order them some money for that purpose they will account for the same.

Washington to Moylan, 14th January, 1780:

His Excellency Governor Trumbull has written to me lately and informed me that the executive of the State of Connecticut are determined to take the most vigorous measures for stopping the intercourse between the inhabitants of that State and the Enemy in New York, and upon Long Island, and has requested me to direct the assistance of the Cavalry should they be found necessary for the more effectual execution of the Law. I have in answer represented to the Governor that the Horse, after a hard Campaign. require as much repose as possible in their Winter Quarters, and have therefore desired him to call for them only in case of emergency. I hope you will not often have occasion to detach the Horse upon business of this kind, but I am to desire you to comply with the requisitions when made. Should they be too frequent we must take an opportunity of remonstrating against the practice.

To which Moylan replied:

COLCHESTER, 1st February, 1780.

I acknowledge the receipt of your Excellency's Letter of the 14th Ulto. by which I am ordered to comply with the requisitions of Governor Trumbull for the assistance of Cavalry. Should they be found necessary to enforce the execution of the Laws to prevent the trade carried on from this State with the enemy, which shall be obeyed.

I am informd that the authority of this town have memorialisd the executive power of the State setting forth the impossibility of their being capable of Quartering the two Regiments in this place during the winter. Indeed they have good reason for it as it will be a difficult matter to find Qrs. even for one, to have them in such a manner as the men can be convenient to attend their horses. The 4th Regiment are at present from absolute necessity dispersed full five miles—and by the inclosed from Major Tallmadge your Excellency will see how the 2d are situated. Had the plan for quartering the two Regiments, which was first formd, taken place they could both have been well accomodated, and I will dare to say at less expence to the publick than they will be at present. I am sorry to find that there is little probability of our having grain to recruit our horses with after a hard duty the last Campaign. One quart p. day is what the Qr. Master tells me can be allowed I mention this that your Excellency may not expect to see the Horse in order for doing duty at the opening of the Campaign. You may be assured, Sir, that nothing on my part shall be wanting to bring them in the best condition which the circumstances will allow of.

Washington at Morristown to Moylan at Colchester, 3d February, 1780:

Since I wrote to you to remove all the Cavalry to Colchester, I have seen a second representation from Mr. Hubbard to the Quarter Master General, in which he seems to confess that they cannot be accommodated with conveniency at Colchester, and wishes Sheldon's Regiment to be left at Weathersfield. I shall therefore leave the Cantonment of the Horse to your Discretion, and have only to recommend to you to keep them as compact as the State of the Forage and Quarters will admit. I should be sorry that there should be any misunderstanding between Governor Trumbull and you, and I think you acted with great prudence in not answering a warm letter from him in the same stile, as you had reason to think he had been unwarrantably prejudiced. You will, upon the whole, find many advantages by cultivating a good understanding with the Civil Authority.

Captains Pike and Craig called upon me for money for the recruiting service. . . . I think you had best turn your attention to reinlisting your old men, and to picking up new Recruits in the Country near the Quarters of your Regiment. This may be done without incurring any extra expense.

John Pierce, Deputy Paymaster-General [appointed June, 1779], writing from Morristown, February 8, 1780, to Thomas Reed at New Windsor, sent him "an account of advances against Colonel Moylan's Regiment which remain unsettled." Mr. Henderson, Paymaster of the Regiment, disputed the justice of the stoppage, as he had mislaid his pay-rolls or left them in Philadelphia. [*Saffel,* 481.]

Moylan to Governor Trumbull:

Major Worthington handed me a Resolve of the General Assembly of this State relative to the two Regiments of Light Dragoons under my command, with your Excellency's order for the removal of the 2d Regiment from Colchester to the severall towns mentiond in the resolve for their Cantonment. I now send the necessary orders to the officers commanding that Regt. that the same may be put in execution as soon as possible.

I am very sorry Sir, that your orders are to have the Regiment so very much dispersed. It will put it out of the power of the officers to pay that attention which is so essentialy necessary for keeping up a proper discipline and will render it impossible for me to fulfil the instructions which I have received from His Excellency the Commander in Chief—from them Sir I take the Liberty to give you the following extracts: "In the Cantonment of the Regiments you will preserve as much compactness as the nature of the place will admit, that by having them all under your own eye, you may be able to keep up good discipline, and prevent dissipation & irregularity.

"You will direct the utmost attention to the Horse, that they may come into the field in the best possible condition for service."

I must plead your Excellency's orders for my not being able to comply with these instructions—had the Regiment been stationed at any one place I would endeavour to fulfill them.

P.S.—The dismounted men of the 2d Regt. are now quartered in Weathersfield. I will be obliged to your Excellency to let me know by the earliest opportunity, whether it is your intention that they should move with the Regiment.

RESOLVE OF CONNECTICUT.

By the Governor & Commander in chief to Stephen Moylan, Esqr., Colonel Comt. of the two Regts. of American Light Dragoons now in this State.

Enclosed is a Resolve of the General Assembly of this State relative to said two Regiments. Pleas to give the necessary orders for the removal of the 2d Regiment call'd Sheldon's, from Colchester to the several towns mentiond in the Resolve, to be canton'd in the proportions and manner therein mentiond.

Given under my hand in the Council Chamber at Hartford, 5th Feby, 1780.

JONTH. TRUMBULL.

Whereas two Regiments of American Light Dragoons are now in this State, in order to be quartered. Wherefore that the same may be distributed in the several towns with the least disadvantage to the Inhabitants, &c. Resolved by this Assembly that the fourth Regiment shall remain at Colchester, where it is now—and the other or 2d Regt. called Sheldon's shall be cantond in the several towns of Farmington, Symsbury, Windsor, Suffield, East Windsor & Enfield, in equal proportions as near as may be, and to be distributed and placed in such parts of said several towns as the Civil Authority & Select Men of such towns shall order & direct.

Moylan reported to Washington:

COLCHESTER, 8 February, 1780.

Inclosed is copy of an act of the Legislature of this State, the Governor's order to me thereon, and my answer thereto, which it is my duty to lay before your Excellency.

Your Letter to me of the 14th Ulto. mentions that I am to comply with the requisitions of the Governor, for the Cavalry, when demanded, which made me not hesitate in obeying his order. Had I not this sanction Sir, was I at liberty to object to the removal of the 2d Regiment? I beg your Excellency's opinion on this point, for my future government.

To which Washington replied:

MORRISTOWN, 16th February, 1780.

With regards to the Act of Assembly of the State of Connecticut, it appears to me founded on a principle which, if extended or carried into a precedent, would be productive of consequences most injurious to the Service. In the present instance, however, there seemed a necessity for complying with it, for the greater ease of the Inhabitants and to prevent the cantonments falling too heavy on any particular place. It is always my wish to accomodate, where no great injuries can result to the service. And I would hope, that notwithstanding the sparse situation into which the cavalry are thrown, the attention of the Officers will provide against the inconveniences apprehended.

REGIMENTAL RETURN WANTED.

On February 15, 1780, Washington to Moylan requesting an exact Return of the number of Non-Commissioned Officers and Privates of your Regiment, designating in a particular manner, how many of them are inlisted for the War, and the different terms of service of the residue, digested in Monthly Columns. You cannot be too expeditious in forwarding me this Return.

At this time there were 130 Pennsylvanians in Moylan's "Regiment."

On 21st Washington directed:

"Should any of the Men belong to the State of Connecticut, you will be pleased to transmit a Return of them immediately to Governor Trumbull. You will notwithstanding this, include them in the Regimental Return which you make to me."

By a return dated Wethersfield, February 22, 1780, made by Major Benjamin Tallmadge of 2d Light Dragoons it appears that in six companies there were but 1 trumpeter, 1 farrier and 5 privates of "Mounted effective dragoons." The reasons given were "Horses much reduced, almost total want of accoutrements, boots and other clothing." Colonel Moylan reported, "Fourth

Regiment returned non-effective for want of breeches, boots, shirts and stockings."

Yet duty had to be performed though the men were not in proper condition. Governor Trumbull from Hartford, on February 19, 1780, sent this order:

AN ORDER FROM GOVERNOR TRUMBULL TO MOYLAN.

At the request of the General Assembly of this State, I have to desire you to order an officer and eighteen mounted Dragoons from the Regiments under your command to march immediately to Greenwich on the Western Coast of this State, for the purpose of protecting the Inhabitants of the adjacent country, and preventing the practice of carrying embargoed provisions to the enemy at New York.

I would wish Lt. Rogers of Sheldon's Regiment to be employed on this service, as being perfectly well acquainted with that part of the country, and, as a Native, particularly interested to be active and vigilant.

Colonel Moylan replied on 21st, saying:

"As you think Natives more particularly interested to be active and vigilant, I have ordered the party from the 2d Regiment."

But three days later he had to report to the Governor that "it was not in his power to furnish the party you made requisition of," as the Adjutant of the Regiment reported "the whole being non-effective." He was "in expectation of some clothing for the 4th Regiment; if it should come timely I can easily equip the number wanting." He advised that "the Assembly ought to appoint some certain plan from which provisions can be drawn for the men and forage for the horses upon this duty."

The same day (24th) Moylan reported to Washington:

That he had received the requisition of the Governor for the eighteen men, to go on the Lines, which I orderd from the 2d Regiment, not doubting but it could be furnished from thence, but such is the state of the two Regiments in respect to cloathing that the party could not be furnished.

Colonel Moylan was then at Colchester, Connecticut.

To Washington he reported from Middletown, 29th February, 1780:

You will see by my Letters, that the Civil powers have taken upon themselves, the cantonment of the 2d Regiment, the 4th are at Colchester, and if we are supplyed with Forage I hope to shew them to advantage, at the opening of the Campaign. I have tryed every method to prevail with the men whose time is near expiring, to reinlist but the desire of seeing home is at present their ruling passion, if I could have granted them that liberty before they march'd so far, it would, I am convinced, have had very good effects but your Excellency's instructions positively forbad it.

ILL CONDITION OF REGIMENTS.

Washington to Moylan from Morristown, 8th March, 1780:

I am exceedingly concerned to see, by the letters which have passed between Governor Trumbull and you, and by the Returns, the ill condition of the 2d and 4th Regiments of Cavalry in respect to Clothing, Arms and accoutrements. I understood that application has been made for the former directly to the Board of War, and I was in hopes that it had been provided. I shall be glad to know what prospect your Regiment has of being supplied and have wrote to Major Tallmadge on the same subject respecting Sheldon's.

A Court Martial is to be held on the 15th April next at Springfield, for the trial of Mr. Tychnor, Deputy Commissary of purchases at Co'os, on sundry charges brought against him by Colonel Hazen. Six Captains and Subs. are to be furnished from the two Regiments of Cavalry: You will therefore be pleased to order that number upon that duty, and direct them to be punctual in their attendance at the time.

Doctor Shippen has summoned you as witness upon his trial which is to be held at this place upon the 14th Instant. After leaving proper directions with the next Officer in command you will repair hither.

TRIAL OF DR. SHIPPEN.

Dr. Shippen was the Director-General of the Military Hospitals. He was charged with selling hospital stores for his own use.

Washington's order did not reach Moylan until March 27th at Middletown, where he had gone on account of the accouchment of Mrs. Moylan, who gave birth to a still-born infant.

Before receiving Washington's order to "repair" to Morristown, Colonel Moylan had sent this declaration:

I do declare upon my sacred honour that in the year 1777 I

went in company with, and by the desire of, Doctor Shippen, the Director-General of the hospitals, to a store, where I tasted five or six pipes of wine; that I recommended them for his own use, as I thought them cheap and good. Given under my hand, this sixth day of March, in Colchester, 1780.

STEPHEN MOYLAN.

The trial adjourned for "a few weeks," wrote Washington to Moylan on April 5th, adding, "It is lucky you did not set out" in obedience to his order of March 8th.

When finally tried, Dr. Shippen was acquitted.

"PREPARING FOR THE SOUTH."

On March 8, 1780, Washington at Morristown, New Jersey, urged the Board of War at Philadelphia to hasten the accoutrements and clothing for the dragoons. The Board replied on March 17th—St. Patrick's Day—that supplies for the cavalry had been ordered sent the camp. Again on March 25th Washington called for saddles for the dragoons "now preparing for an immediate departure for the South."

On the 27th he wrote Colonel Moylan that the Board of War had given directions to provide "uniform clothing for your Regiment" and would "procure caps, leather breeches and boots for the cavalry," and "make provision of swords, pistols and cartouch boxes." Washington had ordered "saddlery." He added:

"Here I must take occasion to enforce a matter which the Board of War have recommended, and which is, to draw no more than the Articles which are indispensably necessary and no more of such Articles than are really deficient. The scantiness of our Stores of every kind and the necessity of retrenching public expenses, by all possible means, makes me hope you will pay the strictest attention to this request."

MOYLAN'S DEAD CHILD.

Moylan to Washington from Middletown, 28th March, 1780:

As your Excellency's Letter of the 8th instant did not reach my hands until yesterday, it rendered my appearance at the trial of Doctor Shippen on the 14th impossible. I will obey your order by sending the six officers to Springfield from the Cavalry to sit on the trial of Mr. Tychner. The only prospect I have of getting cloathing for the 4th Regiment is the promise of the Clothier

General. I have sent Captain Pike near two months past on that business but have not since heard from him, as to the accoutrements I followed the instructions of your Excellency in applying to the D. Q. M. Genl. I know that without these necessary articles as well as some new arms, these two Regiments cannot take the field.

Mrs. Moylan has had the misfortune to be delivered of a dead child which has kept me in this town of late more than I otherwise should be as her situation requires my every attention, I hope it will sufficiently excuse my absence from Colchester. I hear from thence every day—the horses are in very good order.

Should Mrs. Moylan's health & circumstances require it, I will be much obliged by your Excellency's giving me a liberty to accompany her into the Jerseys in April or May. If I should find it detrimental to the service, I shall not make use of it.

On April 1, 1780, Moylan at Middletown sent Washington the "returns for the 2d and 4th Regiments" for March.

On April 5th Washington notified Moylan that Dr. Shippen's trial had been postponed. He would get notice of the next meeting of the Court, "which I imagine will suit that of your attendance upon Mrs. Moylan to Jersey." He sent his own and Mrs. Washington's compliments to Mrs. Moylan "and condolence in your late misfortune."

In a Postscript he added:

P.S.—The Court Martial is adjourned to the 15th May. You will therefore come down about that time, if no material duty in the line of your command should prevent you.

PENNSYLVANIA'S DRAGOONS.

To President Reed of Pennsylvania Colonel Moylan reported from Colchester, April 14th:

The Resolve of Congress passed the 15th March, 1779, respecting the Corps of Light Dragoons, has but very lately come to my knowledge, probably owing to the duty of that part of the Army Which I have the honour to command, being generally employed on the enemies Lines; and of course we are not regularly supplied with the General Orders.

Mr. John Sullivan, mentioned in this Return as belonging to the State, has not been long enough in America to have gained a Settlement in any part of the United States, but as I made him the offer of a Lieutenancy in the Regiment during his Sojourn in Philadelphia. I have Sett him down as appertaining to Pennsylvania, which is agreeable to the Resolve of Congress before mentioned. He is a young gentleman of some fortune, and one that I have every reason to think, will do honor to himself, and to the State of which he is to be a member.

I have great satisfaction in assuring the Legislature of Pennsylvania that no men in the army have done their duty with more alacrity than those in the 4th Regiment of Light Dragoons, who belong to that State, have done.

By the "Return" it was shown that 11 were engaged for the War, 5 until April, 1 until July and 1 to September. Total 18, of whom "only eleven are to be credited as part of the State's quota."

DEPLORABLE CONDITION OF MOYLAN'S MEN.

The deplorable condition of the Dragoons is thus stated to General Washington:

COLCHESTER, 14th April, 1780.

On the first inst. I requested liberty to convey Mrs. Moylan to her friends in Jersey in this or the next month. She is recovering fast from her late indisposition and if I have your permission, I should be glad to set out early in May that I may have sufficient time to return before the Regiment can take the field. I propose taking the tour of the 2d Regiment next week, the 4th are recruiting fast. As to the horses, the men much to be pitied for want of *bread, shirts, boots & stockings* and by much more, for want of *Breeches,* I was informd that it would be deranging the system established for cloathing the Army, to get an order on Boston for the necessary clothing wanting for the 4th Regiment. Colonel Sheldon was not told so, & has got his Regiment comfortable, I am glad he has, but it makes the poor fellows of the 4th feel, for the neglect shewn them, especially as they think, they have at least as much title to attention as any Regiment of horse in the service. There are a few recruits picked up and they tell me 750 dollars is given as a bounty, by Major Lee & others recruiting westward of us, I wish to know from your Excellency whether I am entitled to give that bounty, as I have promisd the utmost given by any officer recruiting for the cavalry. It sickens me to see the 4th Regiment mouldering away. Every day now, carries off some whose times are up, and I have no inducements to offer for encourageing their reinlisting. Badly paid with money that will not purchase an egg in this place; no bread to eat, and seeing themselves and those whom they leave behind almost naked, these are not inducements for continueing in the service.

How the officers manage, is to me inconceivable. I do declare that if I had not drawn upon France for 100 guineas, which by great economy carries me through, I should not be able to get the necessaries of Life. When things are at the worst, they will mend, is the proverb, I hope it will prove true in the present instance. A report prevails that your Excellency is to move Head Qrs. to the Southward. My ambition is to serve where your Excellency immediately commands the army. If it should be true I hope you will not leave me behind you.

This distressful condition of the army owing to the lack of supplies, even "the necessaries of life," was still further made deplorable by reason of the severity of the weather. Washington writing to Lafayette—then in France—on March 18, 1780, from Morristown, stated:

"The oldest people now living in this Country do not remember so hard a winter as the one we are now emerging from," on account of the "extreme cold and deep snows." [*Sparks,* VI, 487.]

On April 18th Washington wrote Moylan:

"There is a quantity of Arms and Accoutrements proper for Cavalry at Springfield. I have directed the Commissary of Military Stores at that place to deliver to you and to Colonel Sheldon for the use of your Regiments, such of the Articles as you may want."

He had written General Schuyler, 22d March, "Our affairs seem to be verging fast to a stagnation in every branch, even provisions."

Moylan to Washington from Hartford, 24th April, 1780:

I came to this place in order to apply to the General Assembly for to advance a sum of money to purchass Forage for the Light Dragoons as what has been collected at Colchester will be consumed in this week, and the inhabitants there, also those in the different towns where the 2d Regiment are placed, absolutely refuse to furnish any upon the credit of the United States. I have expectation that money will be lodged in the hands of the Depy. Qr. Mr. General for this purpose.

I could wish that Lt. Col. Temple was to join the Regiment

before my departure for Morristown in order to attend Doctor Shippen's trial, which I find is adjourned to the 15th May, at which time I hope personally to pay my respects to your Excellency.

I will leave this place to-morrow in order to visit the different Troops of the 2d Regiment and will report the state of that and of the 4th Regt. to your Excellency when I have the honour to see you.

Mrs. Moylan hopes soon to thank you & Mrs. Washington in person for your kind concern on her late misfortune. We propose setting out the 5th of next month at which time I will be the bearer of the Returns for the month of April.

On April 28, 1780, the Supreme Executive Council of Pennsylvania received from Colonel Moylan under date of Colchester, April 14th, a "Return of Officers and Privates of the 4th Regiment of Light Dragoons belonging to the State of Pennsylvania," by which it appears that Moylan was, on June 5, 1776, Colonel by brevet and on 8th June, 1777, in the cavalry from Pennsylvania. [*Col. Rec.*, XII, 332.]

JUNE 1780.

On June 7, 1780, Washington established his headquarters at Springfield, N. J. On the 26th, when at Morristown, he sent orders to General Glover under whose command Moylan's Regiment had been placed (while he was at the home of his wife) to move on to Springfield for the purpose of receiving and forwarding the draft.

On June 10, 1780, Washington at Springfield, New Jersey, directed General Robert Howe, who though of the same name as the British brothers Howe, was a Patriot, to "consider West Point as the capital object of your attention," as there "is a suspicion of something being intended against that post." Clinton's Brigade "may shortly reinforce you. The enemy have a good many cavalry and we have none here. You will despatch immediately Moylan's regiment to join us." [Sparks', VII, 75.]

Washington was moving the army Southward.

Washington to Colonel Stephen Moylan or Officer commanding 4th L. D., Springfield, June 20, 1780:

Should the whole of your Regiment have left King's Ferry, you will be pleased to order back a Commissioned Officer and six Men, with directions to the Officer to remain on this side, and dispatch a Dragoon every morning with a written report of any appearances upon the Water. Should any Vessels heave in sight, he will endeavour to ascertain their number and size, he will come on himself with the last man.

If the whole Regiment should not have come on, you may send these orders to the Officer in the rear.

By Colonel James McHenry, Washington's orders on July 5, 1780, to Moylan were:

His Excellency requests that you will without delay take Post with your Regiment in a situation near the rear of the Army calculated to afford you a sufficiency of Forage. From hence you will detach a commissioned Officer with fifteen Men, which you will be pleased to relieve as often as you may judge proper, and direct them to patrol the Country from the little Falls as far as Acquackenunch Bridge to Hackensack. It is the General's particular recommendation that the patrol frequently change ground nor remain long in a place, never two nights at a time. Major Lee is directed to patrol on your left.

Moylan made reply the same day to which came this answer:

I communicated your answer to His Excellency. He thinks the little Falls too near the Enemy and leaves it to you to take a position within the fork of the River, or on this side, or on the left of the Army as you may find most convenient to the objects which he communicated. You will be pleased to advise the General as soon as you have determined on a position.

BULL'S FERRY.

On July 20, 1780, Washington ordered General Anthony Wayne to "proceed with the first and second Pennsylvania Brigades and Colonel Moylan's regiment of dragoons upon the execution of the business planned."

This was an attack on the block house at Bull's Ferry near Fort Lee on the west bank of the Hudson River.

On July 22, 1780, General Wayne from Totowa reported to General Washington that "the first and second Pennsylvania brigades, with four pieces of artillery belonging to Col. Proctor's regiment and Colonel Moylan's dragoons, took up their line of march on the 20th at three o'clock P.M. and arrived a little in the rear of New Bridge at nine in the evening. We moved again at one in the morning in order to occupy the ground in the vicinity of Fort Lee and the landing opposite King's Bridge by the dawn of day we advanced with . . . Colonel Moylan's horse on the common road. Colonel Moylan with the horse, and a detachment of infantry, remained at the fork of the road leading to Paulus Hook and Bergen, to receive the enemy. . . . Colonel Moylan's horse drove the cattle from Bergen up towards the Liberty Pole whilst detachment of infantry destroyed the sloops and the woodboats at the landing."

Washington reported to President Huntingdon of Congress on July 26, 1780:

Having received information that there were considerable number of cattle and horses in Bergen Neck, within reach of the enemy and having reason to suspect that they meant shortly to draw all supplies of that kind within their lines, I detached Brigadier-General Wayne on the 20th with the first and second Pennsylvania brigades with four pieces of artillery attached to them, and Col. Moylan's Regiment of dragoons to bring them off. I had it also in contemplation to attempt, at the same time, the destruction of a Block house erected at Bull's Ferry, which served the purpose of covering the enemy's wood choppers and giving security to a body of Refugees, by whom it was garrisoned, and who committed depredations upon the well-affected inhabitants for many miles round.

General Wayne having disposed of his troops so as to guard the different landing places on the Bergen shore and having sent down the Cavalry to execute the business of driving off the stock.

While Wayne was attacking the block house, "the dragoons in the meantime drove off the stock which were found in the Neck, the sloops and wood boats in the dock near the block house, were burnt and the few people on board them made prisoners." Wayne lost 15 killed, 49 wounded. [*Pa. Packet,* Aug. 1, 1780.]

Washington to Moylan, Bergen County, 28th July, 1780:

There is a necessity for moving the Army from this Ground to-morrow morning, and as we have not a sufficiency of waggons for the purpose, you will be pleased to divide the whole of your Horse in small parties and send them five or six miles each way to make an impress. Sixty Waggons at least will be wanted, which are to be sent, as they are collected, to the Grand Parade. They must if possible be there some time in the night, as the Troops are to march at three in the Morning. The people may be informed that they will be discharged in three Days and perhaps sooner.

Our move is of the utmost importance and you will for that reason direct the parties to be active in the execution of their business. The people should bring Forage with their waggons if possible.

A DISAGREEABLE CRISIS.

On 30th July, 1780, Washington informed Congress "Of the most disagreeable crisis to which our affairs are brought in the Quartermaster General's department, unless measures are taken immediately to induce General Greene and other principal officers of that department to continue their services, there must of necessity be a total stagnation of military business. We must not only cease from the preparations for the campaign, but in all probability shall be obliged to disperse, if not disband, the army for want of subsistence." [Sparks', VII, 126.]

THE TRAITOR ARNOLD.

On September 27, 1780, his brother, then in Philadelphia, wrote in Spanish to Stephen on business matters but saying, "Yesterday we learned the sorrowful news that Arnold had been declared a traitor and is now in New York." Stephen replied, in French, concerning the business and saying that he had been ill by fever and was yet in such a condition that he could scarcely write. [*MS.*]

Colonel Moylan to President Reed from Camp, October 1:

Major Andre is to be executed this evening at 5 o'clock. What a pity it is not Arnold that is to suffer in his room. His conduct through the examination has been open, candid, manly, and has gained him the esteem of every one. He has been led into the scrape against his judgment, and fortunately for America, by the bad conduct of Arnold in sending him back, was catched. [Reed's *Life of Reed,* II, 276.]

"The man was noble but his last attempt wiped it out and his name remains to ensuing ages abhorred."

On November 17, 1780, Washington sent Moylan instructions:

You will immediately send off all your infirm and reduced Horses to Lancaster in Pennsylvania, at or near which place your Regiment will be cantoned for the Winter.

You will give the Officer who goes with the party directions to deliver to the Deputy Quarter Master General in Lancaster County all the Horses that you are of opinion will not be fit for

Dragoon service another Campaign; the remainder you will have recruited and put in good Order in the course of the Winter.

You will particularly attend to that part of your former Instructions, pointing out the number of Horses to be kept by each Officer according to his Rank, and see that the number is not on any account exceeded. You will have all the Old Accoutrements repaired, as soon as the Men are fixed in Quarters.

As I presume that the State of Pennsylvania is to compleat your Regiment to the establishment of the 3d and 21st October (which you have seen published in the General Orders of the 1st Instant) I would recommend it to you to wait upon the President yourself, or direct an Officer to do it, with an exact state of the Regiment as to Men and Horses, and inform him of the deficiencies in both. You will particularly explain to him, that the times of a great part of your Men expire in the course of the Winter, lest, in making provision for filling up the Corps, he may count upon them.

I do not know the mode that will be fallen upon to procure the Men and Horses: You will therefore govern yourself by the directions which may be given to you by the Executive Authority, in consequence of the measures which may be adopted. Neither do I know whether the State or Continent are to provide your Accoutrements. You will make yourself acquainted with that circumstance, and apply in season to one or the other of those necessary for a full Corps, as it is to be presumed that the State will fully comply with the requisition of Congress.

I shall direct the Officer, with the part of the Regiment which remains here, to repair to Lancaster as soon as the Army quits the Field.

Colonel Timothy Pickering, Q. M. G., to President Reed, 1780:

Col. Moylan's regiment of light dragoons being under marching orders for Winter Quarters, and the upper part of Lancaster county being judged the most eligible for the purpose, I have written to Col. Philip Marsteller, Assistant Quarter Master for that county, to use his endeavours to procure both forage and quarters for them.

Col. Moylan has at present about a hundred and seventy dragoon horses. Many of these are quite unfit for service, and will be turned over to the quarter masters for waggon and riding horses when recruited. Should his regiment be completed he will want a great number of horses to mount his recruits—probably two hundred. These I have not the smallest prospect of being able to purchase. And as the regiment is assigned to the State of Pennsylvania as a part of its quota, I am naturally led to your Excellency as its patron.

Washington had matured an attack on the British in New York at this time. His orders to Colonel Moylan on November 21, 1780, were:

At nine O'clock in the Morning of Friday 24th you will parade with your Regiment at Totawa Bridge, furnished with two days provisions, and you will detach parties towards the New Bridge, and thence upwards to the Bridge near Demarest's and downwards, as far as you think necessary to answer the purpose to secure all the crossing places on the Hackensack, and prevent any person going with intelligence to the Enemy. You will select a sufficient party of the trustiest of your Men to patrol from the Marquis's Old Quarters below the Liberty Pole towards Bergen Town, Bull's ferry, Wehawk, Hobuck &c. A Vidette to be constantly at Bull's ferry and make reports every two hours.

Much depends on the punctual execution of these Orders for which I have intire reliance on your judgment and knowledge of the Country. I confide the rest to your discretion.

P.S.—Van Heer's Corps will join you and take your Orders.

The project, however, failed. Washington wrote that, "circumstances rendered the prosecution of the design inexpedient." [*Ford,* IX, 38.]

CHASTELLUX AND MOYLAN.

At this time the Marquis de Chastellux visited Washington's Camp at Morristown. From his *Travels,* Vol. I, pp. 141-173 this extract relating to Colonel Moylan is taken:

The reader will perceive that it is difficult for me to quit General Washington: let us take our resolution briskly then, and suppose ourselves on the Road. Behold me travelling with Colonel Moylan, whom his Excellency had given me, in spight of myself, as a companion, and whom I should have been glad to have seen at a distance, for one cannot be too much at one's ease in travelling. In such situations, however, we must do the best we can. I began to question him, he to answer me, and the conversation gradually becoming more interesting, I found I had to do with a very gallant and intelligent man, who had lived long in Europe, and who has travelled through the greatest part of America. I found him perfectly polite; for his politeness was not troublesome, and I soon conceived a great friendship for him. Mr. Moylan is an Irish Catholic; one of his brothers is Catholic Bishop of Cork,* he has four others, two of whom are merchants, one at Cadiz, the other at L'Orient; the third is in Ireland with his family; and the fourth is intended for the priesthood. As for himself, he came to settle in America some years ago, where he was at first engaged in commerce; he then served in the Army as Aid de Camp to the General, and has merited the command of the light cavalry. During the war he married the daughter of a rich merchant in the Jerseys, who lived formerly at New York, and who now resides on an estate at a little distance from the road we were to pass the next day. He proposed to me to go and sleep there, or at least to take dinner; I begged to be excused, from the fear of being obliged to pay compliments, of straightening others, or of being myself straightened; he did not insist, so that I pursued my journey, sometimes through fine woods, at others through well cultivated lands, and villages inhabited by Dutch families.

Chastellux and Moylan remained at Morristown over night, "where the goodness of General Washington and the precautions of Colonel Moylan had procured entertainment at the inn of Mr. Arnold."

The next morning they left Morristown.

Some miles from thence, we met a man on horseback, who came to meet Colonel Moylan with a letter from his wife. After reading it, he said to me, with a truly European politeness, that we must always obey the women; that his wife would accept of no excuse, and expected me to dinner; but he assured me that he would take me by a road which should not be a mile out of my way, whilst my people pursued their journey, and went to wait for me at Somerset Court-house. I was now too well acquainted with my Colonel, and too much pleased with him, to refuse this invitation; I followed him, therefore, and after crossing a wood, found myself on a height, the position of which struck me at first sight. I remarked to Colonel Moylan, that I was much mistaken if this ground was not well calculated for an advantageous camp; he replied, that it was precisely that of Middlebrook, where General

---

* He did not become Bishop of Cork until 1787. He was Bishop of Kilkenny in 1780.

Washington had stopped the British in June, 1777, when Sir William Howe was endeavouring to traverse the Jersey to pass the Delaware, and take Philadelphia.

It is here that Colonel Moylan's father-in-law has fitted up a little rural asylum, where his family go to avoid the heats of summer, and where they pass whole nights in listening to the song of the mocking bird for the nightingale does not sing in America.

We soon arrived at Colonel Moylan's, or rather at Colonel Vanhorn's his father-in-law. This manor is in a beautiful situation; it is surrounded by some trees, the approach is decorated with a grass plot, and if it was better taken care of, one would think ones-self in the neighborhood of London, rather than in that of New York. Mr. Vanhorn came to meet me; he is a tall, lusty man, near sixty years of age, but vigorous, hearty, and good humoured; he is called Colonel from the station he held in the Militia, under the English Government. He resigned some time before the war: he was then a merchant and cultivator, passing the winter at New York, and the summer in the country; but since the war he has quitted that town, and retired to his manor, always faithful to his country, without rendering himself odious to the English, with whom he has left two of his sons in the Jamaica trade, but who, if the war continues, are to sell their property and come and live with their father. Nothing can prove more strongly the integrity of his conduct, than the esteem in which he is held by both parties. Situated at ten miles from Staten Island, near to Rariton, Amboy, and Brunswick, he has frequently found himself in the midst of the theatre of war; so that he has sometimes had the Americans with him, sometimes the English. It even happened to him once in the same day, to give a breakfast to Lord Cornwallis and a dinner to General Lincoln. Lord Cornwallis, informed that the latter had slept at Mr. Vanhorn's, came to take him by surprise; but Lincoln, getting intelligence of his design, retired into the woods. Lord Cornwallis, astonished not to find him, asked if the American General was not concealed in his house: "No," replied Mr. Vanhorn, bluntly. "On your honour?" says Cornwallis. "On my honour, and if you doubt it, here are the keys, you may search everywhere." "I shall take your word for it," said Lord Cornwallis, and asked for some breakfast; an hour afterwards he returned to the army. Lincoln, who was concealed at no great distance, immediately returned, and dined quietly with his host.

The acquaintance I made with Mr. Vanhorn being very prompt and Cordial, he conducted me to the parlour, where I found his wife, his three daughters, a young lady of the neighborhood, and two young officers. Mrs. Vanhorn is an old lady, who, from her countenance, her dress, and her deportment, perfectly resembled a picture of Vandyke. She does the honours of the table with exactness, helps every body without saying a word, and the rest of the time is like a family portrait. Her three daughters are not amiss: Mrs. Moylan, the eldest, is six months advanced in her pregnancy; the youngest only twelve years old, but the second is marriageable. She appeared to be on terms of great familiarity with one of the young officers, who was in a very elegant undress, forming a good representation of an agreeable country squire; at table he picked her nuts for her, and often took her hands. I imagined that he was an intended husband; but the other officer, with whom I had the opportunity of conversing as he accompanied us in the evening, told me that he did not believe there was any idea of marriage between them. I mention these trifles only to show the extreme liberty that prevails between the two sexes, as long as they are unmarried. It is no crime for a girl to embrace a young man; it would be a very heinous one for a married woman even to show a desire of pleasing. Mrs. Carter, a handsome young woman, whose husband is concerned in furnishing our army with provisions, and lives at present at Newport, told me, that going down one morning into her husband's office, not much decked out, but in a rather elegant French undress, a farmer of the Massachusetts' State who was there on business, seemed surprised at seeing her, and asked who that young lady was. On being told "Mrs. Carter." "Aye!" said he, loud enough for her to hear him, "A wife and a mother, truly, has no business to be so well dressed."

At three o'clock I got on horseback with Colonel Moylan and Captain Herne, one of the young officers I had dined with. He is in the light cavalry and consequently in Colonel Moylan's regiment.

At four o'clock I set out, after separating, but not without regret, from the good Colonel Moylan.

Colonel Moylan made this report to General Washington then at New Windsor, New York:

PHILADELPHIA, December 7, 1780.

Colo. Temple had waited on the Board of War & the President of this State; shewd them the instructions I received from your

Excellency respecting the 4th Regiment of Light Dragoons. The former could do nothing, the latter has laid the matter before the Assembly, individuals of which inform me they are disposed to do everything in their power for recruiting men and purchassing horses for the Regiment but the finances of the State are in so deplorable a condition that they do not know when they will have the ability to put their will in execution. It is reccomended that I should remain in this place until some plan can be form'd for this purpose.

Washington sent this order to Moylan from New Windsor, 8th December, 1780:

You will be pleased to make to the Board of War as soon as possible, an accurate Return of the Men and Horses in your Regiment, the quantity of Cloathing, and number of Arms, Accoutrements and Furniture of every kind fit for service, and what will be the deficiency, estimating the strength of your Corps at the New Establishment. You will also make a Return of the Articles of Clothing and Accoutrements drawn by the Regiment since the 1st September, 1779. The foregoing are called for by the Board of War, for their government, as they are directed by Congress to make provision for the Cavalry.

HIS BROTHER IN SPAIN.

John Jay, Minister to Spain, writing to Robert Morris from Madrid, December 18, 1780, said:

"When you see Colonel Moylan tell him that his brother is here and very well. We see each other often. He formerly lived at Cadiz—but as Government ordered all the Irish to remove from the seaports, he was obliged with many others to quit it. It is said that their too great attachment to Britain occasioned this Ordinance." [*N. Y. His. Soc. Coll.,* 1878, p. 454]

On New Year's night, 1781, "a most general and unhappy mutiny took place in the Pennsylvania Line," as General Wayne reported the next day from Mount Kemble to Washington, that one-half of the men had taken part. The men angered by long delayed payment and their demand to be discharged as their time of enlistment had passed. "Their general cry is to be discharged and that they will again enlist and fight for America, a few excepted."

On January 4th Colonel Moylan was at Princeton from where he wrote President Reed at Philadelphia:

"I joined General Wayne this day in order to give any assistance that may be in my power. As the enemy will in all probability come out, if the Line will act as they say they will, I shall then be of some service. Should your Excellency think it would be my duty to join my regiment or stay to see the end of this affair pray let me know by bearer. I think my presence at Lancaster may well be dispensed with, I hope you will think so." [*MSS., Pa. His. Soc.*]

As the mutinous soldiers were Pennsylvanians, President Reed went to the scene of action. He wrote to the Committee of Congress on January 9, 1781: "I had the pleasure of meeting General Wayne and Colonels Butler, Stewart and Moylan who have been permitted to come out freely to me. They make such a report of the good temper of the men and their anxiety to see us that I have concluded to go among them. These are the only officers they allow to have communication with them or pass their posts."

The revolt was satisfactorily settled after two British Spies sent to seduce the men to British allegiance had been hung by the revolting soldiers. [*Penna. Arch.*, 2d Series, Vol. XI.]

On January 26, 1781, Colonel Moylan made this report of the officers of the Dragoons who belonged to Pennsylvania:

"Return of the officers of the 4th Regt. Light Dragoons belonging to the State of Pennsylvania: Col. Stephen Moylan, Capt. John Craig, Lt. John Sullivan, Thos. M. McCalla, Surgeon; William Thompson, Riding Master."

The next day a certificate was issued from the War Office to the Supreme Executive Council of Pennsylvania that they were Officers of the Pennsylvania Line.

Preparations were being hastened to carry on the campaign of 1781 in the South where the British forces were being sent, causing trouble and making desolation reign.

Washington, at New Windsor, New York, on February 26th, wrote General St. Clair:

"Congress have determined that the Pennsylvania Line except Moylan's dragoons and the troops upon command to the westward, shall compose part of the Southern army and have directed me to order it to join the army in Virginia by detachments as they may be in readiness to march. It was essential that one of the Brigadiers should proceed with the first detachment. Upon Gen. Wayne devolves the duty." [*St. Clair Papers,* I, 541.]

The Board of War, on March 13, 1781, wrote President Reed of Pennsylvania:

A very considerable Time ago we ordered the Dragoons of Col. Moylan's Regiment to take the Guard at Lancaster of the Magazine & publick stores, & are at a Loss to know why our Orders were not complied with. The Store & Ammunition are by no Means so considerable as the Inhabitants represent but some Attention should be paid to them, & we thought we had provided for their Security by ordering the Guard before mentioned. We shall call upon the Commanding Officer to know the Reasons of his Failure to put our Directions into Execution.

Colonel Moylan, from Lancaster, 22d March, 1781, reported to Washington:

We get some men, but no prospects of horses or accoutrements yet in view. Congress have *resolved* that they shall be got, and there it rests. I will return to Philadelphia to-morrow, and if my public duty will not prevent it, propose bringing Mrs. Moylan from Jersey to this place, where I shall be happy in receiveing your Excellency's commands.

My brother James has sent you a case of Claret which I deliverd to the D. Q. M. G. to be forwarded—he prays your Excellency would pardon the Liberty, and accept of it, as a small mark of the veneration he has for your exalted character. These Sir are his words, which I know to be correspondant to his sentiments.

To which General Washington made reply from New Windsor, 4th April, 1781:

I have written both to the Congress and to the Board of War, and used every Argument to induce a speedy completion of the Regiments of Cavalry. They will be more than ever useful, now the active scene is, in a measure, transferred to the Southward.

You will be good enough, the first opportunity you have of writing your brother in France, to thank him for his present of Claret. It has not yet come to hand.

General St. Clair to President Reed, April 3, 1781:

"I have received information of great uneasiness prevailing among the soldiery, occasioned by the detention of their bounties and the non-payment of the gratuity to the re-enlisted soldiers—unless they are soon made easy on those heads, it is likely to end in general desertion. I must mention also the case of Col. Moylan's regiment. The General expected to march with the detachments; they have not more than fifty horses fit for service, and are in want of every kind of equipment." [*St. Clair Papers,* I, 544.]

On April 6th General St. Clair, to Washington from Philadelphia, wrote:

"That three battalions of 960 men were ready to march in a week to Yorktown. Col. Moylan's Regiment is in such a situation that it must be a considerable time before they can possibly move, having but eighty men and fifty horses fit for service, in want of every equipment and no money in any of the departments to procure them." [*Ibid,* I, 546.]

Washington to Congress on April 8, 1781:

"I wish the march of the Pennsylvania troops could be facilitated and that Moylan's cavalry could be recruited, equipped and marched without delay; for every judicious officer I have conversed with from the southward, and all the representations I received thence, confirm me in the opinion, that great advantages are to be derived from a superior cavalry." [*Sparks',* VIII, 2.]

Washington, from New Windsor, N. Y., 18th April, 1781, wrote to General Greene:

"I have again urged Congress to recruit, equip and forward Moylan's Dragoons to you with despatch." [Ford's: *Writings,* IX, 221.]

On May 16, 1781 General Wayne, at Lancaster, wrote President Reed of Pennsylvania relative to the necessaries wanted by Colonel Moylan, especially cattle, to serve the detachment until it arrived in Virginia.

The same day he wrote the State's Agent, Henry, at Lancaster, for the necessaries actually wanted for fitting out sixty horse of Colonel Moylan's regiment whose services to the southward are of the utmost importance. I wish to have the cavalry in readiness to take up their line of march in two weeks.

Henry replied that it would give him great pleasure if it were in his power to fit out the sixty Light Dragoons, but he had advanced a large sum of money of Pennsylvania to the United States and did not know when it would be repaid. He had some leather belonging to the United States which would furnish leathery accoutrement for the sixty dragoons if the President of the Council would order the leather to be made up, which could be done in two or three weeks—the United States to pay for the workmanship.

On May 19th the Board of War asked President Reed to give the order to provide the accoutrements to Moylan's Horse. All money advanced out of the State treasury be charged to the United States. [*Pa. Ar.,* I, 9, 139.]

That day an "unhappy affair" occurred at the Lancaster Camp which was reported by Colonel Adam Hubley to President Reed, May 21, 1781:

Necessity as well as duty obliges me to give Council a short narrative of an unhappy affair which happened on the 19th instant between the Guards and Col. Moylan's Dragoons. It appears one of the Dragoons for some offence which he had committed was put into the Gaol-house. A rescue by the Dragoons was agreed on. They accordingly assembled armed with Pistols and swords march'd to the Barracks and one more daring than the [others] stept up to the sentinel who previously desired him not to advance or he would put him to death. This however had no effect, he (the Dragoon) first cocking & presenting his loaded Pistol attempted to sieze the sentinel's arms who instantly fired & killed him. The deceas'd's Pistol in his falling went off also & wounded one of the militia men in the thigh.

May 22d, General Weedon to Colonel Grayson: "We shall shortly send sixty of Moylan's horse; also 300 new recruits from Maryland and Delaware." [*Papers Md. Line,* 146.]

LAFAYETTE.

The Marquis de Lafayette in reporting to Washington from Richmond, Va., 24th May, 1781, relating the situation of affairs in that quarter said: "Cavalry is very necessary to us. I wish Lauzun's legion could come. I am sure he will like to serve with me; and as General Greene gave me command of the troops in this State, Lauzun might remain with me in Virginia; if not, Sheldon's dragoons might be sent. As to Moylan, I do not believe he will be ready before a long time." [Sparks' *Corr. Rev.,* III, p. 322.]

The "utmost importance" of even a portion of Moylan's command to reinforce Lafayette in Virginia is made manifest by General Weedon to Moylan from Hunter's Heights, June 8, 1781:

I am just from the Marquis' [Lafayette] camp who labors under every disadvantage for want of horse. He is informed 60 of your regiment is ordered to the Southward and requested me to drop you a line, with his compliments, weil knowing that a knowledge of his situation would be a sufficient inducement to hurry you on.

Indeed he is to be pitied. The enemy have near 400 cavalry; he has only 40 that can be called established dragoons; this superiority of horse gives the enemy a decided advantage and subjects his parties to every evil. In short, if he is not speedily reinforced, they must overrun our country.

Understanding you are in Philadelphia, I refer you to my friend Grayson for news. [*Papers Md. Line,* p. 150.]

CONGRESS RESOLVES.

The United States in Congress Assembled, June 12, 1781,

Resolved, That it be earnestly recommended to the State of Pennsylvania to raise, accoutre and equip Moylan's legionary corps to its full complement the men to be raised for three years unless sooner discharged.

That it also be earnestly recommended to the said State immediately to raise and equip three troops of Militia Cavalry properly officered;

That the said cavalry be marched by detachments as they are compleated to the southern Army and the militia Cavalry discharged in proportion to the numbers enlisted into Moylan's regiment and joining the said army properly armed and accoutred.

Colonel Moylan was at this time at West Point, New York, gathering his men for the southward march. On June 18th he reported to General Washington then at Springfield, New Jersey:

Captain Craig is come with 38 horses and only waits for cloathing of which the men are very destitute, to proceed to Head Qrs. I expect the cloths down from Newboro this evening.

The remainder of the Regiment are coming on and hope they will be at King's ferry or its vicinity the day after to-morrow; if the enemy do not prevent us I will move on with them with all possible dispatch. Should we be interupted in this quarter, I will march them over by way of Fishkil.

Washington writing from New Windsor, 27th June, 1781, to Major Tallmadge of the 2d Regiment of Dragoons:

" Colonel Moylan's Regiment is on its way to join you, which will render the duty easier and your troops there more respectable." [*Sparks' W.*, VI, p. 278.]

A roster of the Field and Staff Officers of Moylan's Regiment, July 3, 1781, may be read in *Pennsylvania Archives*, 5th Series, Vol. III, pp. 835-51, and of Non-Commissioned Officers, 1781-1783.

To Dr. James McHenry, Aide-de-Camp and Secretary to Washington, Colonel Moylan wrote on 5th July, 1780:

" It is his Excellency's desire that I should post my Regiment near the rear of the army at a place calculated to afford a sufficiency of Forage. I have been from the right to the left of the army, in its front and in its rear, and can assure you, that except I was to crowd in upon the ground occupied by Major Lee's Corps, or get upon Pumpton plains I do not know a place where a Regt. of horse could subsist themselves in the Rear or Right. The 4th is at present at the little Falls and shall be glad to know whether I shall move them to, for I assure you I know not, except I get upon the road leading to Morristown. There's good ground about two miles to the Left of Head Quarters."

MOYLAN GOING TO THE SOUTHWARD.

Colonel Stephen Moylan to President Reed, Lancaster, 10th July, 1781:

The Detachment from the 4th Regt. which will leave this to-morrow takes off all the Subaltern officers except two. Recruits are daily coming in, and in a short time, as the Last act for recruiting the Line comes to operate, many more may be expected; it will therefore be absolutely necessary that more Subalterns should be appointed to that Regiment.

President Reed replied to the request for the appointment of the Lieutenants, saying:

Upon looking over the Arrangement printed last Winter we find the number of officers amount to 15, By the Accounts we have of the strength of the Regimt. it is stated to us at 80; We would therefore wish you to consider whether it will not be most conducive to the public Good to defer the appointment of new officers till the Regiment is farther compleated.

MOYLAN " TAKES A POST IN FRONT."

The army moved Southward and by the close of the month were in Virginia.

Colonel James McHenry, Aide-de-Camp to LaFayette and afterwards Secretary of War in the administrations of Washington and Adams, relates under date of July 30, 1781, that "General Wayne and General Morgan are at Good's bridge on the South side of the James River. Col. Moylan and one regiment of light infantry will cross to-day to take a post in front, the militia and the remainder of the infantry on this side." [*Life*, p. 39.]

On October 1, 1781, Moylan's Dragoons were at Williamsburg, Virginia.

YORKTOWN.

From "Camp before York, 8th Octr., 1781," Washington issued this order to Moylan:

There being an absolute necessity for reinforcing General Greene with Cavalry as expeditiously as possible, you will immediately collect all the Men and Horses of the 4th Regiment and report to me the Articles of Clothing of which you stand in need that I may endeavour to furnish you out of a few things (though not of the proper kind for Dragoons) which are coming from the Head of Elk.

In the Order of Battle of Yorktown the first or Right Division was: (1) Pennsylvania Volunteers Battalion of Riflemen under Major William Parr; (2) Fourth Regiment Continental Light Dragoons, Colonel Stephen Moylan. [*Mag. Am. His.*, Oct., 1881.]

So it will be seen Moylan's command had a very honorable, though dangerous position. Lord Cornwallis, besieged at Yorktown, surrendered on October 19th. The captive army moved with grace and precision. Universal silence was observed amidst the vast concourse, and the utmost decency prevailed; exhibiting in demeanor an awful sense of the vicissitudes of human fortune, mingled with commiseration for the unhappy. [Lee's *Memoirs*, 512.]

MOYLAN'S DRAGOONS.

(Supposed to be sung in honor of Moylan's Dragoons, after the surrender of Cornwallis, at Yorktown, in 1781.)

BY THOMAS D'ARCY M'GEE.

Furl up the banner of the brave,
  And bear it gently home,
Through stormy scenes no more to wave,
  For now the calm has come.

Through showering grape and drifting death
  It floated ever true;
And by the signs upon our path,
  Men knew what troop went through.

Our flag first flew o'er Boston free,
  When Graves's fleet groped out:
On Stony Point, reconquered, we
  Unfurled it with a shout;
At Trenton, Monmouth, Germantown,
  Our sabres were not slack;
Like lightning, next, to Charleston
  We scourged the British back.

And here at Yorktown now they yield,
  And our career is o'er.
No more thou'lt flutter o'er the field,
  Flag of the brave!—no more.
The Redcoats yield them to "the Line";
  Both sides have changed their tunes.
To peace the Congress doth incline;
  And so do we Dragoons.

Furl up the banner of the brave,
  And bear it gently home;
No more o'er Moylan's march to wave.
  Lodge it in Moylan's home.
There Butler, Hand and Wayne, perchance,
  May tell of battles brave,
And the old flag on its splintered lance
  Above their heads shall wave.

Hurrah, then, for the Schuylkill side—
  Its pleasant, woody dells!
Old Ulster well may warm with pride
  When each his story tells.
Comrades, farewell! May Heaven bestow
  On you its richest boons!
So let us drink before we go,
  To Moylan's brave Dragoons!

Poets are not historians. What "flag" Moylan's Dragoons had is unknown. Strange as it may now seem the Stars and Stripes was not usually borne by the Army. The only Stars and Stripes carried by a Revolutionary Regiment now preserved is that of the Third Maryland Regiment now at the Capitol at Annapolis.

The poet speaks of "Old Ulster" as if "Butler, Hand and Wayne" and also Moylan were born in that Province of Ireland.

Butler was born in Dublin; Hand in King's County; Wayne in Chester County, Pennsylvania, and Moylan in Cork. Old "Ireland" might the better be used than "Ulster."

On October 26th Washington issued this order to Colonel Moylan:

MOYLAN ORDERED TO JOIN GREENE.

Out of the captured Horses and Accoutrements, and such others as belong to the public, and are to be obtained, you are to equip the first, third and fourth Regiments of Cavalry and Colonel Armand's Corps, and prepare the whole, with as much expedition as possible, for a march to reinforce the Army under the command of Major-General Greene.

On October 31st Moylan was ordered by Washington:

JOINS ST. CLAIR.

In addition to my Orders of the 26th, I have to desire that you will prepare as large a Body of Horse as you possibly can, and join (at such time and place as Major Genl. St. Clair shall appoint) the detachment which he is marching to the Southward. In this I do not mean to include Colonel Armand's Partisan Corps, as it will have a particular place of rendezvous assigned it. But the Infantry of the first, third and fourth Regiments are to be comprehended.

Such Officers belonging to these Regiments respectively, as you shall conceive necessary, may be left to take charge of and forward on the residue of the Corps, the Invalids, Sick and necessaries appertaining to each. A good and active Officer of Rank should have the general direction of the whole to prevent confusion, delay, and that misapplication of time and means, which, unhappily, are but too often met with in our Service.

Though in command of three "Regiments" of Cavalry and of the infantry attached to each, the report of Colonel Moylan shows the skeleton condition of the bodies designated "Regiments" which Moylan was to command in the army of General Greene. He reported to Washington from the Camp, November 1, 1781:

MOYLAN'S "REGIMENT."

It is my duty to lay before your Excellency the situation of my Regiment. It consists of three field officers, six Captains and five Lieutenants. They have 94 men to command.

One field officer, one Capt. & two Lieutenants with 40 men are gone to Carolina, the remainder are taking care of some of those captured horses which will not be fit for service these four months—if ever. Pensilvania, to which the Regt. belongs have hitherto done nothing for it. A Letter from your Excellency to the executive power of that State on this subject must be of great service, and if I was permitted to be the bearer of it, I think the Regt. may yet be on a respectable footing. I have many reasons for requesting this favor, the principal one is, that my health at this time is very bad indeed. A flushed face gives me the appearance of health, whilst an inward fever and an obstinate Dissentry is preying on my vitals. Added to these a total loss of appetite; such is my condition with respect to health. Perhaps the northern climate may restore it. Should that be the case you may be assur'd Sir, I shall loose no time in joining the Southern Army.

Col. Armand has got all the horses but 45. Col. White & Col. Baylor divide the arms & accoutrements, according to the strength of their Regiments. I have sufficient for my men, at Lancaster, which can be got to the detachment long before the horses can be fit to march. I have not yet got the returns; when I do the Qr. Mr. Genl. shall be furnished with them. Pistols & bridles are very defficient, however I think 200 horse can march with General St. Clair.

To which Washington replied the same day:

Your letter expressing your desire to return to Philadelphia with a view of engaging the Executive of Pennsylvania to exert themselves in favour of this part of their Quota.

I cannot conceal from you, Sir, that it is with pain I see my Instructions answered by applications of this kind. The Journey to Philadelphia I am so well persuaded will be fruitless as to the principal object that the trouble and expense of it ought to be avoided. The additional motive of ill health, if your indisposition is of a nature so serious as to incapacitate you for service, is one of those necessities that must be conclusive.

The following day Washington's Secretary wrote Colonel Moylan:

MOYLAN'S ILL HEALTH.

" In consideration of your Health, he consents to your going to Philadelphia, where you are to use your endeavours for placing your Regiment in the best state for Service. If any recommendation from him will be of service, he is willing to second your application, although he has but little hope of success from that Quarter.

"Before your departure the General wishes you to see every arrangement made that is necessary for the Cavalry moving on to the Southward as expeditiously as possible."

To which Moylan answered:

CAMP, November 3, 1781.

Please to make my acknowledgement to his Excellency, for his kind consideration of the state of my health and let him know that I do nòt mean to leave this State, until I find every thing *en train* which can contribute to the forwarding as many horse as possible with General St. Clair. I shall go this day to Williamsburg on my way to Petersburg (the latter place is the rendezvous for the Cavalry). I wish the General would give a Letter to the Executive of Pennsylvania respecting the 4th Regt. L. D., it may be of great service, and can have no ill consequences, I will call at Head Qrs. before I set off, and be glad to take it along with me.

WASHINGTON AND THE FRIENDLY SONS OF ST. PATRICK.

Colonel Moylan came to Philadelphia. He was present at the December meeting of The Friendly Sons of St. Patrick when "His Excellency General Washington was unanimously Adopted a member of this Society," records the Minutes. It was ordered that the Medal of the Society be presented to Washington and that he "and his Suit be invited to an Entertainment to be prepared & given him at the City Tavern on Tuesday, the first of January," to which were invited "the President of the State & of Congress, the Minister of France, Mr. Marbois, Mr. Otto, the Chief Justice, the Speaker of the House of Assembly, Mr. Francisco Rendon, Mr. Holker, Count de la Touche & Count Dillon with all the General Officers that may be in the City."

So on New Year's Day, 1782, the extra meeting was held. The Generals present were Washington, Lincoln, Steuben, Howe, Moultine, Knox, Hand, McIntosh. There were twenty-one guests and thirty-five members present, but Colonel Moylan was not in attendance. He was at the St. Patrick's Day dinner, however.

Where now is the medal of The Friendly Sons which was presented Washington?

When General St. Clair arrived at Washington's camp, General Wayne was ordered, on January 4, 1782, to Georgia, having under him Lieutenant-Colonel White, who had lately joined the army with one hundred Light Horse, the remains of Moylan's regiment of dragoons. Wayne proceeded without delay and in a few days crossed the Savannah River at the Two Sisters Ferry. The immediate object of this motion into Georgia was to protect the country from the incursions of the British at Savannah and to establish the authority of the United States. [Lee's *Memoirs*, p. 539.]

But with the further movements of this army and its remnant of Moylan's Dragoons we are not concerned, as Moylan was not in command of the men who had so long followed him.

On June 13, 1782, General Gist was appointed by General Greene commander of all the Light Troops. The cavalry of Lee's Legion, "the Third Regiment" and Moylan's Fourth Regiment being placed in command of Colonel Baylor. [*Ibid*, 552.]

In October, 1780, Captain B. Edgar Joel made charge against Major-General Robert Howe, who had been engaged in the defense of Georgia against the British, that he had sacrificed Savannah, December 29, 1778. On October 18th the Board of War referred the charges to General Washington, then at Passaic Falls, New Jersey. The operations of the army debarred trial until a Court-Martial, over which Baron Steuben presided, was held at Philadelphia on December 6, 1781. Colonel Moylan was a member of the Court. He served until January 3, 1782.

General Howe was "acquitted with the highest honor."

There were two Howes in the British service in America—Sir William Howe and Viscount Richard Howe.

After the surrender of Cornwallis at Yorktown, military operations practically ceased on both sides. The projected campaign in the South did not take place, so Colonel Moylan was not called into service. He remained at Philadelphia. The annexed document attests his presence there while Washington was at Newburg, New York, pending the consideration of Articles of Peace. Secretary Tilghman was the writer on 20th September, 1782:

Sergeant Morris of your Regiment has applied to His Excellency for a discharge upon procuring another Man in his Room. This is a practice that His Excellency would not wish to tolerate, but as Morris seemed much discontented, and is of ability to do mischief in the Corps, by stirring them up, by a frequent recapitulation of their grievances, he thinks it best to get rid of him upon the terms he offers. You will therefore give Orders to the Commanding Officer at the place of rendezvous to discharge him upon procuring an able bodied and otherwise good Man in his stead.

Though the army was inactive awaiting the terms of peace, consolidation arrangements were being perfected so as to make the force more concentrated for service, if needed.

Colonel Moylan's concern at the possibility of having his command taken from him by the measures projected is shown in his letter to Washington:

MOYLAN "A MAN WHO HAS SACRIFICED EVERYTHING FOR THE SERVICES OF HIS COUNTRY."

PHILADELPHIA, December 15, 1782.

The Minister of War informing me he intended paying a visit to your Excellency on the subject of the new arrangement, I take the liberty of communicating what past between us, respecting the 4th Regiment of Light Dragoons.

By the returns it appears, that Regt. is reduced to two Troops, one mounted, the other not. These are to be commanded by one Field Officer two Captains & the proper number of Subalterns. General Lincoln says that he supposes a Major will be appointed to this command, in which case, after eight years service I shall be laid aside. This command tho inadequate to my rank would be more agreeable—for when the Army is in the field, that objection would be removed, and being the oldest in the Line of Cavalry I am thought to believe, that the senior in each Line have it in their option, to continue in service or not, as they choose.

In this Line there is not a Field Officer (myself excepted) that belongs to the State of Pensilvania, tho I dare say, they will be perfectly satisfied with what your Excellency may determine on; it is very natural to suppose that they would prefer an officer of their State, to that of another to command their own Troops.

I have made application to the Executive power of this State, who promised me all their influence towards completing the 4th Regt.

I mentioned this circumstance to the Minister of War; he told me Congress did not wish to increase the Cavalry—which has put a stop to any farther proceedings of mine in that business.

When I entered the service—which was early in the first year of the war—I did it with a firm determination of prosecuting it to the end. I had made up my mind, and my affairs for that purpose. I have shared its fatigues, its dangers and its pleasures, with your Excellency ever since—should I be now left out, I shall be very much disapointed and very much distressed. General Lincoln informs me, it will depend upon you to officer the Corps. I am sure you will do it, in the manner which will appear to your Excellency most consistant with rectitude. From the polite and friendly attention I have allways experienced from you—I have expectations, that you will retain in the Army a man who has sacrificed everything for the service of his Country.

Should there be anything inconsistant with the new plan of arrangement by my remaining in the Line I now serve in, it may be in your Excellency's power to find other employment for me, if such should offer near your own person, it would be very pleasing to me.

To this Washington made reply on December 25, 1782:

WASHINGTON'S CHRISTMAS MESSAGE TO MOYLAN.

I have been favoured with the receipt of your Letter of the 15th Instant, concerning the reduction of your Regiment, and have conversed with the Secretary at War on the subject. That Gentleman (on whose determination will depend the particular modes to be adopted in the reformation of the Lines not serving immediately under my Orders) has gone to the Eastward, without deciding any thing positively as to the reduction of the Cavalry. However, it appears to me, to be his prevailing opinion that as your Legion could not remain entire, but must be reduced to the broken part of a Corps, it would be best to annex the company of Infantry as a Flank Company to one of the three Regiments of Pennsylvania Line, still allowing the Officers and Men the pay and Emoluments they formerly enjoyed. Should this be the case, or at any rate, I imagine the remainder of the Legion will soon be too small a command for even a single Field Officer of any grade.

But as nothing can be ultimately determined upon before the return of General Lincoln, and as some event may happen or intelligence arrive in the mean time, which will lead to an alteration in the Plan of arrangement, I thought it expedient to take the first opportunity to advise you of these particulars and to assure you (whatever the circumstances of the public or the service may eventually require) of my unalterable Esteem and Regard.

General St. Clair to President Dickinson of Pennsylvania:

PHILADELPHIA, February 10, 1783.

I enclose an arrangement of the fourth Regiment of light Dragoons and Colonel Moylan's Return. The Return did not come to my Hands untill this Day, for want of which, and some Knowledge of the Intentions of Major Fauntleroy the arrangement has been delayed, it must however be supposed to have taken place upon the first of January, and I have dated it accordingly—that Corps will consist of two Troops one mounted and one dismounted.

General St. Clair continued that Major Fauntleroy if "in service on the first of January, is of course continued in the command of the two troops," but as he had been long absent and "it is even doubtful whether he has not resigned, and had been guilty of neglect if not contempt in having on a former occasion gone away contrary to orders," General St. Clair concluded the office should be considered vacant.

So "after eight years service" General Moylan was "laid aside."

MUTINY.

On April 19, 1783, Washington issued an order declaring "a cessation of hostilities" and advising "the Patriot Army to retire from the military theatre with the same approbation of angels and men which has crowned all their former virtuous actions." The army was disbanded eight years to a day after the Revolution had begun. But the army had not been paid. Some parts not for years. Washington's suggestion that those enlisted for the war should be allowed to retain their arms and accoutrements was sanctioned by Congress. In May it allowed him to grant furloughs or discharges while awaiting settlement of accounts and securing of funds. Within two months most of the soldiers had gone home and without being paid. A body of Pennsylvanians, enlisted as late as November, 1782, and stationed at Lancaster and Philadelphia, mutinied and marched to the State House in Philadelphia where Congress was in session demanding that justice be done them within "twenty minutes," threatening to "let in these injured soldiers upon you." They seized the magazine. The mutiny continued a week. An account of it may be read in Scharf & Westcott's *History of Phila-*

delphia, Vol. I; Hatch's *Administration of the Revolutionary Army*, Chap. IX, and other publications.

Congress removed to Princeton on account of this mutiny.

MOYLAN'S MUTINEERS.

Our concern is with Moylan's Dragoons who were concerned in this revolt. We have seen how Colonel Moylan procured a Lieutenancy for John Sullivan, newly arrived, and had him credited to Pennsylvania's quota. He was one of the Committee of the Mutineers threatening Congress. A rumor that Washington, with a body of three-year men, retained until the British would evacuate New York, and the New Jersey Militia, were coming to the protection of Congress caused the mutineers to submit. Their committeemen made flight. Sullivan and a fellow-member, Captain Henry Carberry of the Eleventh Pennsylvania Regiment reached Chester and from there wrote the third: "Consult your own safety; we cannot get to you." They embarked for England. [Hatch: *Adm. Rev. Army*, 186.]

SULLIVAN'S DEFENSE.

On June 26, 1783, Sullivan, from Chester, wrote Colonel Moylan:

"Do not blame me. The success of every enterprise generally demonstrates it right or wrong. Had not the soldiers betrayed us, we should have carried our point or perrished in the attempt."

On June 30th he wrote from "The Cape":

"If a consciousness of Rectitude can be a consolation to men in Adversity, be assured our Spirits are far superior to our circumstances and I am Confident that none but persons accustomed to judge of things by the event will reprobate our conduct. The little prospect of succeeding in such an attempt and the difficulties we had to encounter were so great, that [not] to have been confident of success would have proved us to be as destitute of prescience as of common sense. But a series of injuries and the incessant indignities we experienced were our sole inducements for prosecuting the plan at all risks. I am not ignorant of the sentiments of the men in power in respect to the army and the ideas they entertain of the passiveness of the officers. Those circumstances determined us to convince them we had a just and right sense of our wrongs and were not callous to ill treatment. It —— little what appearences —— may assume to veil Injustice but its a duty encumbent on honest men, by investigating the principles of —— policy not to submit to the imposition. Actuated by the present motives of patriotism and disinterestedness, I abandoned my dearest concessions (?) at a tender age to fight under American colours at a critical period and when affairs were equally balanced. For my conduct in the army either as a Soldier or a gentleman, I appeal to the officers in general and you in particular to decide on. I flatter myself you will not suppose that my attachment for this Country is diminished in the most trivial degree. I ever had an innate affection for America and were she on the verge of ruin, I would come and perish with her. Let what bad men there are at the helm of Government observe from this instance how dangerous it is to drive men of honor to desperation. The person concerned with me is Capt. Carberry. This Young Gentleman served with eclat in the Army, bled and spent a pretty fortune in the service of his country. I have requested Capt. Than [?] to be punctual in paying what triffling debts I owe and have requested your Brother John to draw on my father if necessary.

"We are wafted along by a gentle and generous gale and possess the most perfect tranquility of mind conscious of no unworthy actions, all we regret is failing in a noble attempt.

"'And more true joy Marcellus exiled feels,
Than Cæsar with a Senate at his heels.'

"Adieu, my dear Colonel, be convinced I have just sense and want words to express my gratitude for the peculiar mark of favor and affection you have confered on me."

These letters of Lieutenant Sullivan are "copies from memory" among the Papers of Ben Franklin, in the American Philosophical Society, Philadelphia.

How did they get there? Why?

Though Sullivan and Carberry had escaped, the other officers concerned were Court-martialed but acquitted. Several of the soldiers were convicted and sentenced to death but when ready for execution were reprieved.

BRIGADIER-GENERAL.

On November 3, 1783, Congress *Resolved* that Colonel Stephen Moylan be promoted to the rank of Brigadier by brevet.

The Roster of "Moylan's Dragoons" at this time may be read in *Pennsylvania in the Revolution*, Vol. II, p. 140.

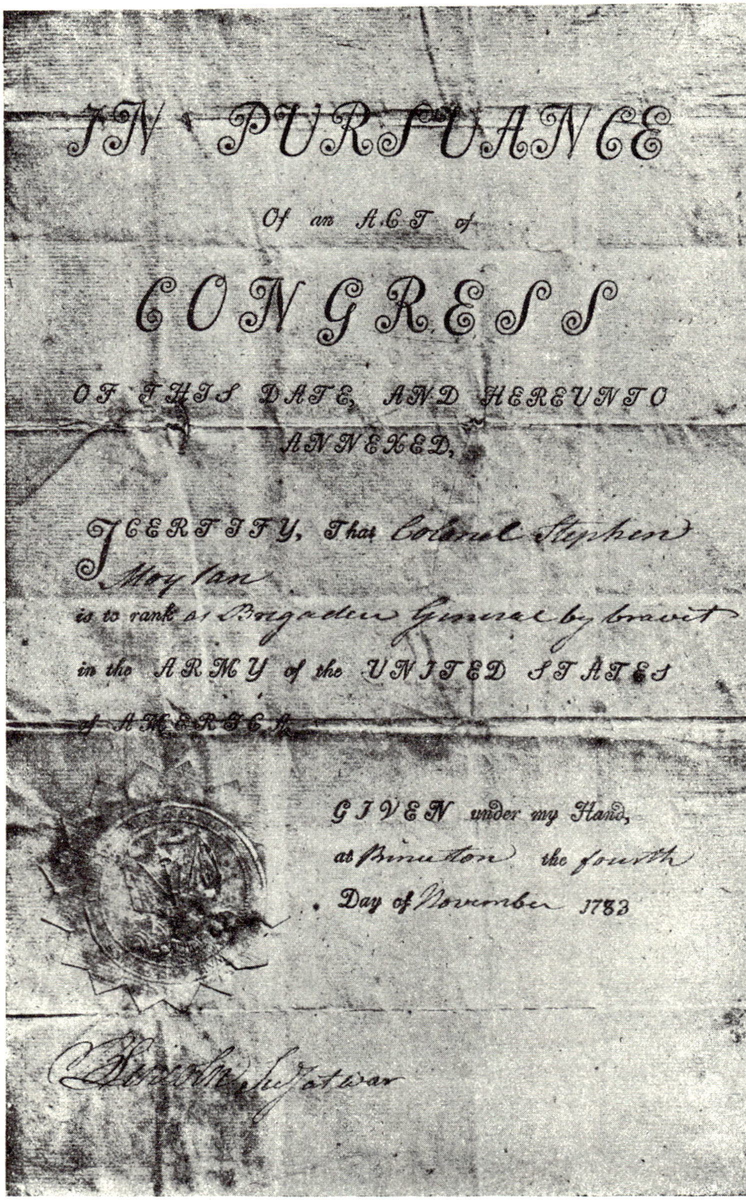
IN PURSUANCE

Of an ACT of

CONGRESS

OF THIS DATE, AND HEREUNTO ANNEXED,

I CERTIFY, That Colonel Stephen Moylan is to rank as Brigadier General by brevet in the ARMY of the UNITED STATES of AMERICA.

GIVEN under my Hand, at Princeton the fourth Day of November 1783

B Lincoln Secy at war

GENERAL MOYLAN'S COMMISSION

VISITS GENERAL WASHINGTON.

In the *Washington Manuscripts* in the Library of Congress is a letter from General Moylan, dated Alexandria, May 16, 1785, saying:

"On looking over the money you favored me with, I found a quarter of a dollar over, which Col. FitzGerald will be kind enough to return to you.

"Permit me to return you my sincere thanks for the polite attentions which Mrs. Moylan and myself received from you & your good Lady during our agreeable sojourn at Mount Vernon. You may be assured it will be long remembered with pleasure."

In 1792 General Moylan was a resident of Goshen Township, Chester County, Pa., on a farm near West Chester.

On April 7, 1792, he was appointed Register and Recorder of Chester County to succeed Persifer Frazier, deceased. He held these offices until December 13, 1793, when he was succeeded by Colonel John Hannum.

On March 20, 1793, Colonel Hannum gave ground upon which St. Agnes's Church, West Chester, is built. General Moylan of Goshen Township was one of the Trustees of the church.

MAJOR-GENERAL OF MILITIA.

In May, 1793, Governor Mifflin of Pennsylvania appointed General Moylan "Major-General of the Division composed of Chester and Delaware Counties." He accepted it at West Chester, May 25th, "with more pleasure as it is a further mark of your friendly attention to me," he wrote the Governor.

OFFERED THE MARSHALSHIP.

Thomas Jefferson, Secretary of State under President Washington, sent the following communication to General Moylan:

PHILADELPHIA, September 13, 1793.

SIR:—

The President, on his departure, left in my hands a commission for a Marshal of this district with a blank for the name to be inserted. It was his wish that your's should be inserted if you should think the office would suit you. I must ask the favor of you to say whether you would accept of the commission, and to do it in a letter to Mr. Benjamin Bankson at my office, as I set out for Virginia within two or three days. Should you decline it I must still ask you to notify it to him, that he may proceed to follow the instructions given him in that case. The office will be vacant on the 20th inst. by the resignation of Col. Biddle, and I can with truth express the satisfaction it would give me personally to have it filled again by a person to whose merits I am less a stranger than to his person. I am with great respect, Sir

Your most obedt. servt.,

TH. JEFFERSON.

Genl. Moylan.

[*Jefferson Papers,* Series 1, Vol. 5, p. 378.]

Mr. Bankson, on September 24th, wrote Moylan:

Not hearing from you am fearful lest the letter may have miscarried. I have therefore judged it expedient to forward this by an Express, as it is of importance that your acceptance or non-acceptance should be known as soon as possible. Please advise me of this by the return of the Express. [*Ibid,* 388.]

[*Jefferson Papers,* Series 1, Vol. V, p. 388.]

General Moylan must have declined but no "letter of declination" is to be found in the Department of State. David Lenox was later offered the position. [*Ibid,* 389.]

On December 3d President Washington appointed General Moylan Commissioner of Loans.

HIS COMMISSION.

This is a copy of the Commission of General Moylan:

TREASURY DEPARTMENT,

December 9, 1793.

SIR:—

The President of the United States having been pleased to appoint you to the office of Commissioner of Loans in the State of Pennsylvania, you will herewith receive your Commission. I request that you will as speedily as possible enter upon the execution of the office.

I am, Sir, with esteem, Your obedient Servant,

A. HAMILTON.

THE LAST PRESIDENT OF THE FRIENDLY SONS.

On St. Patrick's Day, 1796, the Friendly Sons of St. Patrick of Philadelphia met at the house of Samuel Richardet. Twenty members were present. "General Stephen Moylan was unanimously elected President and Thomas FitzSimons, Esq., was elected Vice-President," is the record and the last known Minute of the Society. But that the Friendly Sons continued to exist is proven by the will of Michael Morgan O'Brien, made September 2, 1803, wherein he bequeathed to his nephew, James Roland, residing on the Island of Dominica in the West Indies a "gold medal which was struck for the members of a Society known by the name of the Friendly Sons of St. Patrick and of which Society General Moylan is President for the present year."

When the Society became extinct is not known. Its heir and successor by merging and possession of the original records of the Society is "The Friendly Sons of St. Patrick for the Relief of Emigrants from Ireland." It was organized in 1793 under the title of The Hibernian Society for the Relief of Emigrants from Ireland, but a few years ago altered its title to the present. It preserves the best traditions of the old Society and amply fulfills the mission of its own foundation, the relief of needy emigrants from Ireland. While its quarterly meetings maintain that spirit of fellowship the foundation of the organization of 1771 whereby those of Irish birth or descent, irrespective of political or religious beliefs unite in good will and with brotherly regard for each other, welded into "one harmonious whole," to uphold the honor of the Race and to perpetuate the memory of all of the old land who have been helpful in the material or social building of our State and Nation. May it continue so to do.

MOYLAN'S RESIDENCE AND OFFICE.

In 1796 General Moylan moved to Northeast Corner of Fourth and Walnut. It had been occupied by Mrs. Payne, with whom James Madison, while a Representative in Congress from Virginia, had lodged. He married Dolly, the daughter of his landlady and so seems to have become manager of the property. It was by him leased to General Moylan who, as tenant, made repairs, deducting expense from rental, to which Mr. Madison objected.

On May 9, 1796 General Moylan wrote Mr. Madison that "the

HOME AND OFFICE OF GENERAL MOYLAN, 1796-1810

room now occupied as an office has undergone no alteration" and if "applied to any other purpose it would need repairs."

HORACE BINNEY.

In a deposition made in 1860, Horace Binney, the distinguished Philadelphia lawyer, declared:

"In the spring of the year 1806 I began to occupy a house on the south side of Walnut St., the first door east of Fourth St. At that time Colonel Stephen Moylan lived nearly opposite to my residence at the North East corner of Fourth and Walnut Streets. He was then Commissioner of Loans of the United States and kept his office in the same house and was frequently called General Moylan. I knew him pretty well."

DEATH OF GENERAL MOYLAN.

On April 13, 1811, General Moylan died. He was buried the next day. The *American Daily Advertiser* of Tuesday, April 16, 1811, had this obituary:

"Died on Saturday morning last, in the 74th year of his age, after a lingering illness, General Stephen Moylan of this City, Commissioner of Loans for the City of Philadelphia. He served with distinction in the American army during the whole of the Revolutionary War, and few of his illustrious associates enjoyed a larger share of the favor and friendship of the Commander-in-Chief, than which a more decisive proof could not be adduced of the elevation of his character and the merit of his services. General Moylan displayed, uniformly, in his domestic and social relations those virtues of the heart which shed most lustre and happiness over private life. The singular tenderness of his nature, the active benevolence of his feelings, the candour and uprightness and generosity of his disposition, the mildness and urbanity of his manners, attached to him by the strongest ties of affection and respect not only the members of his own family, but all those who formed the numerous circle of his friends. His remains were on Sunday interred in the burial ground of the Catholic Church of St. Mary's, and attended by his brethren of the Cincinnati and the body of his private relatives and particular acquaintances."

General Moylan's death has heretofore been stated as of April 11th, and such is the record at the Treasury Department at Washington. But by the *Advertiser's* obituary it will be seen that "Saturday last" was April 13th.

He made no will. His estate, valued at $800, was administered to by his half-brother, Jasper Moylan [*Adm. Book K,* p. 424] to whom letters of Administration were given May 11th, Jared Ingersoll and John Hallowell, Counsellors-at-Law, being the sureties.

Though General Moylan was buried in St. Mary's graveyard, the location of his grave is not known even to his descendants. The compiler in 1876 gave special attention to the quest but without satisfactory results. The General's daughter was then alive and a resident of Baltimore. On interview with her by Mr. Patrick Moylan a resident of that city, but not a relation, Mrs. Lansdale stated that a head and footstone had been erected at the General's grave but the stones had been removed when a general order for the removal of all stones had been given. There is an error about the removal of all stones, as many stones yet standing show. Whether so marked or not the grave is not now known.

Until 1810 General Moylan lived at Northeast Corner of Fourth and Walnut. In that year he moved to No. 230 [now 618] Spruce Street where he died. April 13, 1811.

The Public Ledger of Philadelphia, August 16, 1908, gave an illustration of the house, saying: " The dwelling having been modernized looks very little like the house in which the soldier died. The principal difference is to be found in the entrance. Originally there was what was regarded as an ornate doorway of wood. The eaves and cornice bore traces of the colonial adaptation of the Greek ornamentation." The house was newly built and General Moylan, most probably, was its first occupant. The *Ledger* continues:

" The war history of General Moylan was not so dramatic or spectacular as that of some of the more popular heroes, but he was regarded as one of the officers upon whom reliance could be put, both as a man and as a soldier."

A writer in Dawson's *Historical Magazine,* July, 1861, said: " General Moylan was emphatically a gentleman of the old school; he was remarkable for his hospitality. Having two daughters, one of whom was very fascinating, his house attracted many young persons."

There is no known portrait of General Moylan. Family tradition is that one was in the Peale Museum when destroyed by fire.

Previous to 1860 a claim for money due General Moylan by the Government was presented in several Sessions of Congress.

A Bill to reimburse his heirs passed one House more than once but never both. Finally the effort was abandoned.

John Pope Hodnett in speech before Senate Committee on Labor and Education, May 1886, said:

> "There goes Moylan on his prancing steed,
> Always ready whenever the need;
> With his prancing chargers and his green plume,
> Driving the enemy to eternal doom."

Poets are not Historians. Moylan did not wear "a green plume."

Moylan, Delaware County, Pennsylvania, is named in honor of the General.

Such is a compact and yet very complete recital of the career of General Stephen Moylan, Washington's friend and compatriot who sacrificed all for his country. To those of his race and Faith the record herewith presented, mainly from official sources, is commended as an inspirative source for presenting him in spirit-arousing sentiments to American citizens, little aware of the extent and import of the services of this Irish and Catholic native of Cork in old and famed Ireland, which has given to our Country so many of its illustrious men.

That but his name is known to his own is a discredit which it is hoped this account of his career will remove. His services were most helpful in winning the Liberty and Independence of our Country.

The Friendly Sons of St. Patrick of Philadelphia might well erect a memorial of so illustrious an American as the first and last President of the original Society.

It is an obligation of paramount importance that acknowledgment be recorded of the aid given to the compiler by Mr. Charles H. Walsh of Washington, D. C., a great-grandson of Jasper Moylan. The copies of great bulk of the documents used in this work which are now in the Library of Congress were given us by Mr. Walsh.

To General Moylan's great-grandson, J. Moylan Lansdale, Esq., of Philadelphia, our appreciation of his help is also freely acknowledged.

THE MOYLANS IN THE REVOLUTION.

In addition to General Stephen Moylan his brothers James and John and his half-brother Jasper are worthy of recognition as Irish Catholics who did efficient service during the American Revolution.

JAMES MOYLAN COMMERCIAL AGENT AT L'ORIENT.

James was a resident of Philadelphia as early as 1771. On June 17th of that year he was present at the quarterly dinner of The Friendly Sons of St. Patrick as a "Visitor" and at that meeting was elected a member. This indicates his recent arrival in the City.

On St. Patrick's Day, 1775, he is recorded as "Beyond Sea." He had gone to France. Here he became helpful to the American cause and when Franklin, Deane and Lee were there as Commissioners of the United States, Lee "thought him a capable and deserving man" to act as Commercial Agent at L'Orient.

Moylan formed a partnership with Gourlade a French merchant. This firm acted as Prize Agents for American vessels and also fitted out the "Ranger" under John Paul Jones, and the "Alliance" under Captains Pierre Landais and John Barry as well as other American cruisers and supply vessels.

He it was who obtained the French vessel the "Duras" for John Paul Jones after the "Ranger" had been taken from his command. The "Duras" had made three voyages to India, "therefore," wrote Moylan to Jones, November 10, 1778, "I cannot recommend her to you for a lasting ship," but it was the only vessel then obtainable. So Jones replied, "She must be ours." She was fitted out at the expense of the King of France, her name changed to the "Bonne Homme Richard" in compliment to Benjamin Franklin. In her Jones commanding an expedition of four French vessels and the "Alliance," an American commanded by the Frenchman Landais had his famous battle with the "Serapis" which he captured and to which he transferred the crew of the "Bonne Homme Richard" which soon sank.

On March 22, 1781, General Moylan wrote Washington from Lancaster, Pa.: "My brother James has sent you a case of claret. He prays your Excellency would pardon the liberty and accept it as a small mark of the veneration he had for your exalted character.

These are his own words which I know correspond to his sentiments." The American translator of the *Travels of the Marquis Chastellux* says James Moylan "was singularly useful in the year 1777, by managing a treaty between the American Commissioners and the Farmers General of France, for an annual supply of tobacco from America, which he concluded, during Lord Stormont's residence at the Court of France, and many months previous to the open rupture with that Court. I speak of this with personal knowledge of the fact, nor was it so secret as to have escaped the English Ambassador, or the vigilant Mr. Forth. There could not be a more direct attack on England and English claims, than this transaction, which must have had the sanction of the French government, yet England was lulled to sleep by her Ministers, or rather was so infatuated as to shut her ears against the most interesting truths. I could say much more on this subject, but why enter into discussions which have long ceased to be either seasonable or useful? England was, literally, in the case of the Quos Deus vult perdere."

The compiler hopes in the near future to present a detailed account of the career of this Revolutionary worthy whose services, though in France, were as efficacious in supporting the American cause as those of his military brother, Stephen, on the battlefields of America.

JOHN MOYLAN, CLOTHIER-GENERAL OF WASHINGTON'S ARMY.

John had been a merchant at Cadiz before coming to America. On March 22, 1781, he was appointed Clothier-General to the Army then at Morristown, New Jersey.

It was a most important charge for at that time the financial condition of the United States was in a most deplorable state and the money valueless, so much that even yet it is a byword: "Not worth a continental."

Yet the Clothier-General had to financier and to procure supplies. An extended account of his services it is intended to present. His letter to General Washington on his election to the Presidency may be now fitly presented:

PHILADELPHIA, 19th June, 1789.

Sir:—I can no longer resist the Impulse of my feeling in thus congratulating you & America on the proof they have lately given of their gratitude & discernment by appointing you once more to preside over their Interests, & in thus testifying at the same time my regret at having been prevented by absence in joining with my fellow citizens in the general Demonstrations of Joy manifested on the occasion as you & Mrs. Washington passed thro this city. May America long enjoy the advantages which her choice seems so happily to presage! May you, Sir, live to witness & enjoy the blessings expected from an Administration begun under such favorable auspices. May you long live as happy as you are beloved is my sincere & fervent prayer! I dare hope, Sir, that you will excuse on the motive which induced it the freedom of this address, which has at present no other view than that of paying a small portion of the Debt of Gratitude, Respect & Attachment I owe you, sentiment with which I shall be over proud to acknowledge myself.

Yr. Excellency's Very Obedient & most Humble Servant,

John Moylan.

His Excellency,
The President of the United States.

ENSIGN JASPER MOYLAN.

Jasper Moylan, half-brother of Stephen and James and John, was an Ensign in Philadelphia Associators of Pennsylvania Militia. He arrived from Spain in 1781 and on October 1, 1781, took the Oath of Allegiance to Pennsylvania and is recorded as " late from the Kingdom of Spain; student at law."

He was a member of The Friendly Sons of St. Patrick and a resident of Philadelphia for many years. An extended memoir may later be presented.